I0076402

AI AWARENESS SERIES

AI in Manufacturing

Robert Thornfield-Wells

© 2025 Robert Thornfield-Wells

All rights reserved.

No part of this publication may be reproduced, stored in a retrieval system, or transmitted in any form or by any means—

electronic, mechanical, photocopying, recording, or otherwise— without the prior written permission of the author, except for brief quotations used in scholarly reviews or analysis.

This publication is designed to provide accurate information on the subject matter covered.

It is sold with the understanding that the author is not engaged in rendering legal or other professional services.

Any resemblance to real organizations or individuals is entirely coincidental.

Contents

Introduction

Manufacturing stands at a critical inflection point. After decades of incremental automation and digital transformation, artificial intelligence has emerged as the catalyst for a fundamental reimagining of how we design, produce, and deliver products. This transformation extends far beyond simple automation—it represents a shift in the very nature of industrial operations, where machines don't just execute tasks but learn, adapt, and optimize in ways that were unimaginable just a few years ago.

The convergence of several technological advances has made this moment possible. Sensors have become ubiquitous and affordable, generating vast streams of operational data. Cloud computing provides the computational infrastructure to process this information at scale. Machine learning algorithms have matured to the point where they can extract meaningful patterns from complex industrial datasets. Together, these technologies create an environment where AI can deliver tangible value across every aspect of manufacturing operations.

Yet the path to AI-enabled manufacturing is neither straightforward nor uniform. Organizations face critical decisions about where to invest, how to integrate new technologies with existing systems, and how to prepare their workforce for fundamental changes in how work gets done. The complexity of modern supply chains, the demands of global competition, and the imperative for sustainable operations all add layers of challenge to an already complex transformation.

Introduction

This book provides a practical roadmap for navigating these challenges. Rather than dwelling on theoretical possibilities or distant futures, we focus on proven applications and actionable strategies that manufacturers can implement today. Each chapter examines specific use cases where AI is already delivering measurable results—from predictive maintenance systems that prevent costly equipment failures to quality control automation that catches defects humans might miss, from supply chain optimization that reduces waste to energy management systems that balance productivity with sustainability.

The future of manufacturing will be defined by those who successfully blend advanced technology with operational excellence. AI provides unprecedented tools for optimization, prediction, and automation, but these tools must be wielded with skill and purpose. The organizations that master this balance—maintaining focus on fundamental manufacturing principles while embracing technological innovation—will define the next era of industrial excellence.

Chapter 1: Data Strategies and Infrastructure

Sensors serve as the primary interface between the physical world and AI systems, converting real-world phenomena into digital data streams. They form the foundation for data collection by capturing temperature, pressure, motion, visual, and countless other environmental variables that AI models require for training and inference. The continuous nature of sensor data streams ensures that AI systems receive up-to-date information for reliable monitoring and decision-making. This real-time data flow is particularly crucial for automation systems in industrial environments, where sensors enable predictive maintenance, quality control, and process optimization, ultimately improving operational efficiency and reducing downtime.

The integration of IoT devices with industrial systems creates a seamless data ecosystem that transforms traditional manufacturing and operational environments. This integration enables continuous data exchange between sensors, controllers, and analytical systems, significantly improving operational efficiency through real-time insights. However, this connectivity introduces compatibility challenges as organizations must ensure different device protocols and communication standards work harmoniously together. Latency becomes a critical concern when systems require immediate responses. Additionally, the expanded attack surface created by numerous connected devices necessitates robust security frameworks to protect against cyber threats and maintain system integrity.

Chapter 1: Data Strategies and Infrastructure

Managing sensor data at scale presents several interconnected
challenges that organizations must address strategically. Data storage
challenges arise from the sheer volume of continuous sensor streams,
requiring scalable and cost-effective storage solutions that can handle
petabytes of information. Transmission bottlenecks occur when high-
volume sensor data overwhelms network infrastructure, causing delays
that can impact real-time decision-making. Ensuring data quality
becomes increasingly complex at scale, as organizations must
implement validation and cleaning processes to maintain accuracy
across millions of data points. Finally, real-time processing needs
demand sophisticated infrastructure capable of analyzing incoming
data streams instantly to provide timely insights for critical operations.

Data lake architectures represent a paradigm shift from traditional
database systems by providing centralized repositories that can store
vast amounts of both structured and unstructured data in their native
formats. Unlike rigid database schemas, data lakes employ flexible
schema designs that allow for adaptive data organization, supporting
multiple AI workloads without requiring extensive data transformation
upfront. This flexibility enables data scientists and engineers to explore
datasets in their raw form, facilitating discovery of unexpected patterns
and relationships. Data lakes particularly excel at supporting AI
workloads by providing the diverse, large-scale datasets necessary for
training sophisticated machine learning models across various domains
and applications.

Edge computing transforms data processing by bringing
computational power closer to data sources, dramatically reducing
latency and bandwidth consumption while enabling faster system

4

responses. This proximity-based approach is essential for real-time AI applications where millisecond delays can impact safety or performance, such as autonomous vehicles making split-second navigation decisions or industrial systems responding to equipment failures. In industrial automation, edge computing enables on-site data processing that improves operational efficiency and safety by providing immediate feedback without relying on cloud connectivity. The autonomous vehicle industry particularly benefits from edge computing's ability to process sensor data locally for critical navigation and collision avoidance decisions.

The combination of data lakes with edge analytics creates a powerful hybrid processing architecture that leverages the strengths of both centralized and distributed computing. Edge analytics provides immediate local processing capabilities, enabling rapid responses to time-sensitive situations while reducing data transmission requirements. Meanwhile, data lakes serve as comprehensive repositories for long-term storage and deep analytical processing across multiple data sources. This enhanced responsiveness from local processing complements the comprehensive analysis capabilities of centralized data lakes, creating a balanced system that can handle both immediate operational needs and strategic business intelligence requirements through aggregated multi-source data analysis.

Understanding the fundamental differences between real-time and batch processing is crucial for designing effective AI systems. Real-time processing handles data instantly as it arrives, enabling immediate analysis and decision-making essential for time-critical applications. This approach prioritizes speed and responsiveness but often requires more computational resources and complex infrastructure. Batch processing, conversely, handles large datasets at scheduled intervals, optimizing for efficiency and throughput rather than speed. The choice between these approaches depends heavily on the specific AI application's latency tolerance and throughput requirements, with many modern systems employing hybrid approaches that combine both methodologies strategically.

Several AI applications require real-time insights to function effectively and safely. Fraud detection systems must process transactions instantly to identify and prevent fraudulent activities before financial losses occur, protecting both institutions and consumers. Predictive maintenance applications analyze equipment sensor data in real-time to identify potential failures before they cause costly downtime, enabling proactive maintenance scheduling. Autonomous systems, whether vehicles, drones, or industrial robots, rely entirely on real-time data processing to make split-second decisions that ensure safe and efficient operation. These applications demonstrate how real-time processing capabilities directly impact safety, financial protection, and operational continuity.

Balancing latency, scalability, and cost represents one of the most challenging aspects of AI infrastructure design. Real-time systems demand minimal latency, often prioritizing processing speed over computational complexity and cost considerations, leading to higher infrastructure expenses. Scalability challenges arise when systems must accommodate growing data volumes and user loads while maintaining performance standards, often requiring architectural redesign and significant resource investment. Cost trade-offs become evident when comparing infrastructure requirements between batch and real-time processing, with real-time systems typically requiring more expensive hardware and software solutions. Effective architectural planning must carefully balance these three factors to meet system requirements while maintaining budget constraints and future growth capacity.

Accurate data labeling forms the cornerstone of effective AI model development and deployment. High-quality labels enable AI models to learn meaningful patterns from training data, directly impacting overall model performance and reliability. When labels accurately represent the underlying data characteristics, models can generalize better to new, unseen data and provide more trustworthy predictions. Conversely, poor or incorrect labeling introduces bias and significantly decreases model accuracy, potentially leading to unreliable or harmful outputs in production environments. Data labeling represents a critical step in AI development that demands careful attention to quality and consistency, as the accuracy of labels directly determines the effectiveness of the resulting AI system.

Data governance and compliance principles provide the framework for responsible AI development and deployment. Ensuring data quality through governance policies guarantees that AI systems receive reliable, accurate input data, leading to trustworthy outcomes and consistent performance. Protecting data privacy requires implementing strict safeguards to prevent unauthorized access to sensitive information throughout the AI pipeline. Regulatory compliance frameworks ensure adherence to evolving data protection laws and industry standards, protecting organizations from legal risks while maintaining operational legitimacy. Supporting ethical AI use through

7

proper governance promotes responsible handling of AI-generated insights and decisions, fostering trust among stakeholders and maintaining social responsibility standards.

Automated tools revolutionize data labeling and governance processes by significantly improving efficiency and reducing manual workload burdens. Automation streamlines labeling workflows through machine learning-assisted annotation, active learning techniques, and semi-supervised approaches that accelerate the creation of high-quality training datasets. Error reduction becomes achievable through automated governance tools that implement consistent validation rules, detect anomalies, and maintain data quality standards across large datasets. Compliance enforcement relies on techniques such as automated audit trails, data versioning systems, and policy monitoring tools that ensure continuous adherence to governance standards and regulatory requirements without requiring constant manual oversight.

AI data pipelines face increasingly sophisticated threats that can compromise system integrity and expose sensitive information. Data poisoning attacks represent a particularly insidious threat where attackers intentionally corrupt training data, causing AI models to produce inaccurate or harmful outputs that may not be immediately detected. Model inversion threats exploit vulnerabilities in deployed AI systems, allowing attackers to reverse-engineer models and extract sensitive training data, potentially exposing private information about individuals or organizations. Unauthorized access risks encompass various attack vectors targeting AI infrastructure, from credential theft to system vulnerabilities, all of which can compromise model integrity and expose confidential business or personal information.

8

Chapter 1: Data Strategies and Infrastructure

Implementing comprehensive security best practices protects AI infrastructures against evolving cyber threats. Strong authentication mechanisms, including multi-factor authentication and role-based access controls, verify user identities and prevent unauthorized system access, forming the first line of defense. Encryption techniques protect data both in transit and at rest, ensuring confidentiality and integrity throughout the AI pipeline, from data collection through model deployment. Network segmentation isolates AI infrastructure components, limiting the potential impact of security breaches and preventing lateral movement by attackers. Continuous monitoring systems detect suspicious activities and provide real-time alerts, enabling rapid response to potential security incidents before they can cause significant damage.

Regulatory compliance in AI security serves dual purposes of legal protection and trust building. Following data protection laws such as GDPR, CCPA, and industry-specific regulations safeguards AI systems against legal risks while maintaining required regulatory standards that evolve with technological advancement. Compliance frameworks provide structured approaches to handling sensitive data, implementing privacy controls, and maintaining audit trails that demonstrate responsible AI practices. Building trust through regulatory compliance fosters user confidence by ensuring ethical and secure AI practices that protect individual privacy and organizational data. This trust foundation becomes increasingly important as AI systems handle more sensitive applications and gain broader societal adoption.

Chapter 1: Data Strategies and Infrastructure

Strong data strategies and robust infrastructure form the foundation for building effective AI systems that can scale and adapt to evolving business needs. Understanding and utilizing sensor data enables organizations to unlock powerful AI insights and applications that drive operational efficiency and competitive advantage. Integrating data lakes with edge computing creates hybrid architectures that enhance processing efficiency and scalability while managing costs effectively. Finally, ensuring data quality through proper governance and securing AI workflows through comprehensive cybersecurity measures unlocks the full potential of AI systems while maintaining trust and compliance.

Chapter 2: Predictive Maintenance

Predictive maintenance represents a paradigm shift from reactive to proactive equipment management. At its core, this approach relies on three fundamental pillars. First, condition-monitoring tools serve as the eyes and ears of your operation, continuously collecting real-time data to assess equipment health and detect the earliest signs of potential failure. Second, data analytics transforms this raw information into actionable insights, processing vast amounts of collected data to predict potential failures and optimize maintenance schedules with remarkable precision. Finally, the ultimate goal is reducing downtime through timely repairs that minimize unplanned interruptions, dramatically improving operational efficiency and reducing costly emergency maintenance situations.

Modern predictive maintenance offers substantial advantages, including extended asset lifespan and significantly enhanced operational efficiency across industrial applications. However, implementation presents notable challenges that organizations must navigate carefully. Data quality remains the most critical obstacle, as high-quality, accurate data is essential for reliable predictions, yet maintaining this data integrity proves consistently challenging in real-world environments. Integration complexities arise when incorporating predictive maintenance systems with existing infrastructure, requiring careful technical coordination and system compatibility considerations. Additionally, implementation costs demand significant upfront investment, impacting budget allocation and long-term planning strategies, though the return on investment typically justifies these initial expenditures.

Four key technologies form the technological foundation enabling modern predictive maintenance success. IoT sensors represent the data collection layer, continuously gathering real-time information from equipment to enable early detection of potential failures across all operational parameters. Machine learning algorithms serve as the analytical engine, processing data patterns to predict maintenance needs and prevent unexpected breakdowns with increasing accuracy over time. Cloud computing provides the essential infrastructure, offering scalable storage and processing power necessary to handle the massive volumes of maintenance data generated by modern industrial operations. Finally, big data analytics interprets these vast datasets to uncover hidden insights, enabling the development of optimized maintenance strategies that maximize equipment reliability and operational efficiency.

Anomaly detection employs four primary techniques to identify operational irregularities before they become critical failures. Statistical analysis serves as the foundation, helping identify deviations from normal operation by analyzing data distributions, trends, and statistical patterns that indicate emerging problems. Thresholding provides a straightforward approach, detecting anomalies by establishing operational parameter limits beyond which automated alerts are triggered, enabling rapid response to concerning conditions. Pattern recognition offers sophisticated detection capabilities, identifying unusual operational patterns by comparing current data streams with established baseline patterns of normal equipment behavior. Unsupervised learning algorithms provide the most advanced detection methods, discovering anomalies without requiring pre-labeled training data by automatically identifying hidden patterns and relationships in operational data.

Anomaly detection leverages three distinct algorithmic approaches, each offering unique advantages for different operational contexts. Statistical models provide classical foundational methods, utilizing well-established algorithms for data analysis and anomaly detection that have proven reliable across numerous industrial applications. These time-tested approaches offer transparency and interpretability that many organizations value. Machine learning techniques represent the next evolution, improving anomaly detection accuracy by learning complex patterns from historical data and continuously adapting to changing operational conditions over extended periods. Deep learning approaches utilize sophisticated neural network architectures to enhance detection accuracy and adaptability, particularly excelling in complex anomaly detection tasks involving high-dimensional data, multiple variables, and subtle pattern recognition requirements.

Successful anomaly detection implementation requires addressing several critical challenges while following established best practices. Handling imbalanced datasets proves crucial for improving model accuracy and reducing prediction bias, as normal operations typically vastly outnumber anomalous events in historical data. Reducing false positive rates enhances system reliability and maintains user trust, preventing alert fatigue and ensuring maintenance teams respond appropriately to genuine concerns. Algorithm integration focuses on seamlessly incorporating detection systems into existing operational infrastructure, ensuring compatibility, scalability, and minimal disruption to ongoing operations. Best practices emphasize continuous model evaluation and collaboration with domain experts, maintaining system effectiveness through regular performance assessment, model updating, and incorporating practical operational knowledge to refine detection accuracy and relevance.

Chapter 2: Predictive Maintenance

Equipment failure mode prediction employs three complementary methodological approaches to forecast potential equipment failures. Physics-based modeling utilizes fundamental physical laws and engineering principles to simulate equipment behavior, providing theoretical foundations for understanding how components degrade and fail under various operational conditions. Data-driven approaches leverage historical operational data and statistical analysis techniques to identify patterns and correlations that indicate developing failure modes, utilizing real-world performance information to predict future problems. Hybrid models combine the strengths of both physics-based and data-driven techniques, creating more comprehensive and accurate failure prediction systems that benefit from theoretical understanding while incorporating actual operational experience and performance data.

Effective failure mode prediction relies on diverse data sources and sophisticated analysis methodologies to achieve accurate results. Failure predictions utilize varied data sources including sensor readings, maintenance logs, operational parameters, environmental conditions, and historical failure records to create comprehensive equipment profiles. Statistical analysis methods process this multifaceted data to detect subtle patterns indicating potential failures, identifying correlations and trends that might not be apparent through simple observation. Machine learning algorithms analyze increasingly complex datasets to develop accurate failure prediction models, continuously improving their predictive capabilities through exposure to new operational data and feedback from actual failure events, ultimately creating robust systems capable of anticipating equipment problems with remarkable precision.

Industry case studies demonstrate the transformative impact of failure mode prediction across various sectors. Predictive models successfully identify potential failure modes before they manifest as actual problems, significantly minimizing unexpected downtime while improving overall operational efficiency and equipment reliability. This proactive approach enables maintenance teams to address developing issues during planned maintenance windows rather than responding to emergency failures. Downtime reduction represents one of the most significant benefits, as failure mode prediction dramatically reduces unexpected equipment interruptions, saving substantial costs while increasing manufacturing productivity and operational continuity. Maintenance cost savings result from accurate failure predictions enabling targeted, condition-based maintenance strategies, significantly lowering overall maintenance expenses across industries by eliminating unnecessary preventive maintenance while preventing costly emergency repairs.

Understanding vibration and acoustic signatures forms the foundation of condition-based monitoring for rotating equipment and machinery systems. Each machine generates distinctive vibration and sound patterns that serve as unique operational fingerprints, reflecting the specific mechanical characteristics, operating conditions, and current health status of individual equipment units. These signatures remain remarkably consistent during normal operation, making them reliable indicators of equipment condition. Deviation detection becomes possible by monitoring variations from established normal vibration or acoustic signatures, which can indicate potential mechanical faults, component wear, misalignment, imbalance, or other developing issues long before they result in catastrophic failures, enabling proactive maintenance intervention.

Modern sensor technologies enable precise vibration and acoustic data collection essential for effective condition monitoring programs. Advanced sensor types, particularly accelerometers and microphones, serve as key instruments for capturing vibration and acoustic data with high accuracy and reliability across various frequency ranges and operating conditions. These sensors can detect subtle changes in equipment behavior that indicate developing problems. Condition monitoring data collected by these sophisticated sensors provides vital information for continuously monitoring machinery and equipment condition, enabling maintenance teams to track equipment health trends over time, establish baseline performance parameters, and identify deviations that warrant further investigation or immediate maintenance action.

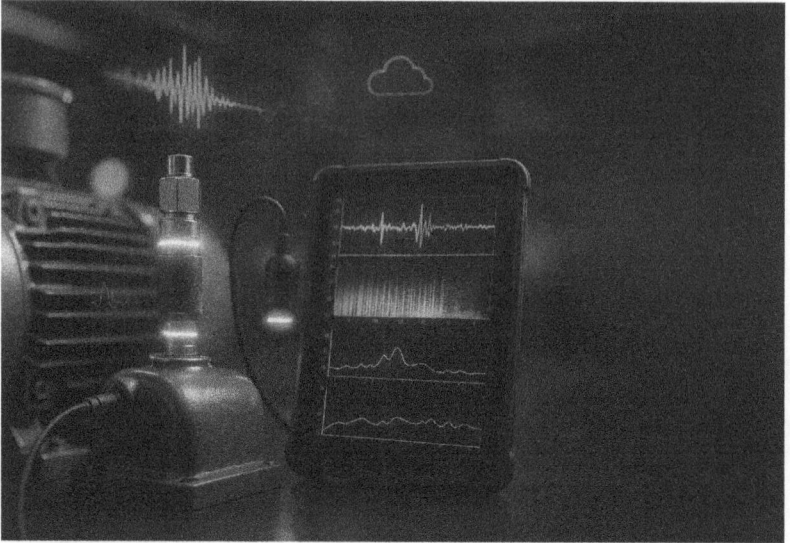

Three analytical techniques transform raw sensor data into actionable predictive insights for maintenance planning. Frequency analysis helps identify repetitive patterns in sensor data, revealing characteristic frequencies associated with specific mechanical components and potential failure modes, enabling maintenance teams to anticipate requirements effectively by recognizing frequency signatures of developing problems. Wavelet transforms provide sophisticated signal processing capabilities, analyzing sensor signals simultaneously across multiple time and frequency scales for detailed feature extraction and early fault detection, particularly useful for identifying transient events and non-stationary signal characteristics. Machine learning models utilize extracted features from frequency and wavelet analysis to predict maintenance needs and optimize operational efficiency, continuously improving their accuracy through experience with equipment behavior patterns.

Digital twins represent a revolutionary approach to equipment monitoring and maintenance planning through virtual modeling. These dynamic digital models create comprehensive, real-time digital representations of physical systems, incorporating geometric, operational, and behavioral characteristics that mirror actual equipment performance. The simulation capabilities of digital twins enable detailed analysis of system behavior under various operating conditions, allowing engineers to test scenarios, predict outcomes, and optimize performance without impacting actual operations. Monitoring and optimization functions enable digital twins to continuously track asset performance, identifying opportunities to enhance efficiency, reduce energy consumption, and extend equipment lifespan through data-driven insights and predictive recommendations.

Digital twin integration with predictive maintenance systems creates powerful synergies that enhance maintenance effectiveness. Sensor

data integration ensures digital twins continuously collect and process real-time sensor information, providing accurate system monitoring and analysis capabilities that reflect current equipment conditions and operational parameters. Advanced predictive algorithms analyze this integrated data to forecast potential failures and maintenance needs, utilizing both historical trends and real-time conditions to generate increasingly accurate predictions. Proactive maintenance becomes possible through insights generated by digital twins, enabling maintenance teams to schedule interventions before problems occur, significantly reducing unexpected downtime while improving overall operational efficiency and equipment reliability.

Virtual modeling through digital twins provides three significant benefits for maintenance planning and operational optimization. Scenario testing capabilities allow maintenance teams to evaluate multiple maintenance strategies virtually, testing different approaches to find the most effective solutions without disrupting actual operations or risking equipment damage. This enables optimization of maintenance procedures and resource allocation. Failure forecasting helps maintenance planners anticipate potential equipment problems, enabling proactive maintenance scheduling and preparation, avoiding unexpected downtime while ensuring necessary parts, tools, and personnel are available when needed. Optimized maintenance scheduling utilizes virtual model insights to reduce maintenance costs and minimize equipment downtime, balancing maintenance frequency with operational requirements to achieve optimal equipment availability and performance.

Three strategic approaches optimize maintenance schedules to maximize equipment availability while minimizing costs. Condition-based scheduling represents a fundamental shift from time-based to health-based maintenance, scheduling maintenance activities based on actual equipment condition rather than predetermined intervals, preventing failures while reducing unnecessary maintenance and associated downtime. Risk-based prioritization ensures maintenance resources focus on the most critical needs, prioritizing tasks by assessing potential risks to operations, safety, and business continuity, ensuring high-impact maintenance receives priority attention. Dynamic real-time adjustment enables maintenance schedules to respond to changing conditions, utilizing real-time operational data to optimize resource allocation, accommodate urgent needs, and adapt to operational changes while maintaining overall maintenance effectiveness.

Chapter 2: Predictive Maintenance

Predictive analytics plays a central role in optimizing maintenance scheduling through data-driven decision making. Forecasting equipment health represents a core capability, as predictive analytics estimate current and future equipment condition to prevent unexpected failures and reduce costly unplanned downtime through early intervention. These forecasts enable maintenance teams to plan proactively rather than reactively. Optimizing maintenance timing utilizes analytical insights to schedule maintenance activities at optimal times, improving operational efficiency while saving costs through better resource utilization, reduced overtime requirements, and minimized production disruptions. This approach ensures maintenance occurs when most beneficial for both equipment health and operational continuity, maximizing the value of maintenance investments.

Scheduling optimization delivers measurable improvements in operational performance across multiple dimensions. Reduced unplanned downtime results from optimized scheduling that minimizes unexpected interruptions, maintaining continuous operations while reducing costly emergency repairs and production losses. This predictive approach enables maintenance teams to address developing issues during planned maintenance windows. Enhanced resource allocation emerges through efficient scheduling that enables better utilization of labor, equipment, and materials, maximizing productivity while reducing waste and unnecessary costs associated with poor planning or reactive maintenance approaches. Improved operational productivity represents the cumulative benefit, as overall productivity increases when maintenance tasks are scheduled optimally, reducing delays, minimizing workflow disruptions, and creating more predictable operating conditions that support consistent production performance.

The benefits of predictive maintenance are clear: improved equipment reliability, reduced operational costs, and enhanced overall efficiency across industrial operations. Sensor technologies provide the critical foundation, collecting real-time data essential for accurate maintenance predictions and enabling the transition from reactive to proactive maintenance strategies. Digital twins represent the future of maintenance planning, simulating assets to predict failures and optimize operations proactively, creating virtual testing environments that enhance decision-making capabilities. Together, these technologies and approaches create comprehensive maintenance strategies that maximize equipment availability, minimize costs, and support continuous operational improvement in modern industrial environments.

Chapter 3: Quality Control Automation

Quality control automation represents a paradigm shift from manual inspection to intelligent, technology-driven processes. Robotic systems eliminate human variability, performing precise, repeatable inspections at superhuman speeds while maintaining consistent accuracy standards. Advanced sensor networks provide continuous monitoring capabilities, detecting anomalies in real-time before they become costly defects. Software integration creates seamless data flow between inspection points and manufacturing execution systems, enabling immediate corrective actions. This trinity of robotics, sensors, and software creates a robust foundation for modern quality assurance. The integration reduces inspection time by up to 80% while improving detection rates significantly compared to traditional manual methods.

Implementing quality control automation delivers substantial benefits while presenting significant challenges. Accuracy improvements stem from eliminating human error and fatigue factors, while operational costs decrease through reduced labor requirements and fewer defective products reaching customers. However, initial capital investment can be substantial, often requiring 18-24 months for ROI realization. System integration complexity increases with existing legacy equipment, requiring careful planning and phased implementation. Perhaps most critically, success depends on skilled personnel who understand both traditional quality principles and modern automation technologies. Organizations must invest in comprehensive training programs to bridge this knowledge gap and ensure sustainable automation success.

Successful automation integration requires careful consideration of three critical factors. Seamless process integration ensures automation enhances rather than disrupts existing workflows, requiring detailed process mapping and stakeholder engagement. Data management coordination becomes paramount as automated systems generate massive datasets requiring real-time analysis and historical trending capabilities. Effective real-time feedback loops enable immediate process adjustments when deviations occur, preventing defective product propagation. Integration success depends on viewing automation as a system-wide enhancement rather than isolated point solutions. Organizations achieving best results implement automation incrementally, validating each phase before expanding scope. This approach minimizes risk while building internal expertise and stakeholder confidence.

Computer vision transforms surface defect detection through four fundamental principles. High-resolution image acquisition captures microscopic details invisible to human inspectors, using specialized cameras operating at speeds exceeding 1000 frames per second. Pattern recognition algorithms identify normal surface characteristics, flagging any deviations as potential defects with remarkable consistency. Edge detection techniques locate boundaries and discontinuities that indicate cracks, scratches, or dimensional variations. Machine learning integration enables systems to improve detection accuracy over time, learning from both correct identifications and false positives. These technologies work synergistically, creating inspection systems capable of detecting defects smaller than human hair width while maintaining production line speeds.

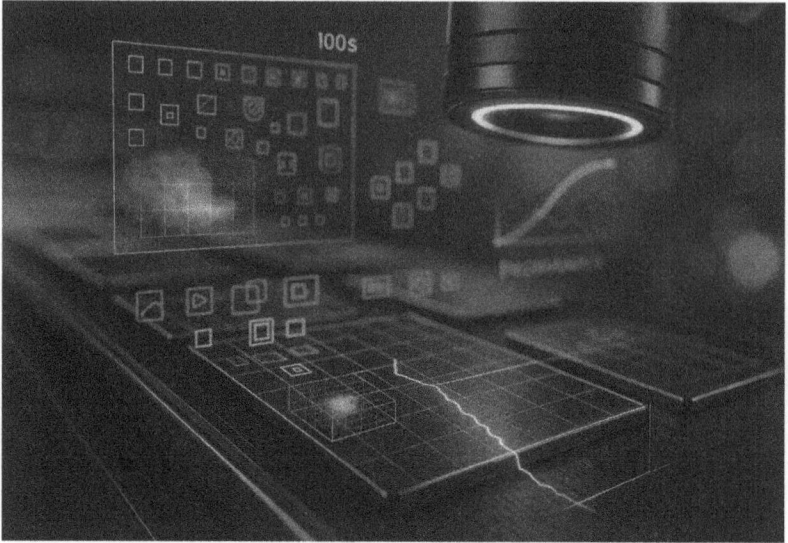

Computer vision systems excel at detecting four primary surface defect categories. Scratch detection analyzes surface texture variations using advanced lighting techniques that highlight microscopic imperfections invisible under standard illumination. Dent identification employs depth mapping and contour analysis, measuring surface topology changes with sub-millimeter precision. Discoloration and contamination detection utilizes sophisticated color analysis algorithms that identify subtle variations in hue, saturation, and luminosity values. Crack detection combines edge detection with pattern analysis, identifying linear discontinuities regardless of orientation or size. Modern systems can simultaneously monitor all defect types, providing comprehensive surface quality assessment. Implementation success requires calibrating detection sensitivity to balance thoroughness with production speed requirements.

Chapter 3: Quality Control Automation

Real-world computer vision implementations demonstrate significant quality improvements across diverse industries. Automotive manufacturers report 95% reduction in surface defect escapes while increasing inspection speeds by 300%. Electronics industry applications achieve microsecond inspection cycles, enabling 100% component verification without production bottlenecks. Packaging operations benefit from simultaneous label verification, seal integrity checking, and contamination detection in single inspection stations. Success factors include proper lighting design, camera positioning optimization, and algorithm training with representative defect samples. Return on investment typically occurs within 12-18 months through reduced warranty claims, improved customer satisfaction, and decreased manual inspection costs. These implementations showcase computer vision's transformative potential across manufacturing sectors.

Statistical Process Control provides the foundation for maintaining consistent manufacturing quality through continuous monitoring and analysis. Traditional SPC applies statistical methods to detect process variations before they produce defective products, using control charts to visualize process stability over time. Variation detection capabilities enable operators to distinguish between normal process fluctuation and abnormal trends requiring intervention. Consistency assurance comes through systematic monitoring of critical process parameters, ensuring output remains within specified tolerance ranges. SPC transforms manufacturing from reactive quality inspection to proactive process control, reducing defect rates while improving overall equipment effectiveness. Modern implementations integrate with automated data collection systems, providing real-time insights previously impossible with manual data recording.

Artificial intelligence revolutionizes traditional SPC by processing massive datasets beyond human analytical capabilities. AI techniques identify complex patterns spanning multiple process variables simultaneously, uncovering relationships invisible to conventional statistical methods. Early deviation prediction enables intervention before process drift produces defective products, potentially reducing scrap rates by 60-80%. Machine learning algorithms continuously refine prediction accuracy using historical process data and defect correlation analysis. Proactive quality control emerges when AI systems automatically adjust process parameters to maintain optimal conditions. This transformation from reactive to predictive quality management represents a fundamental shift in manufacturing philosophy, enabling unprecedented quality consistency while reducing operational costs through optimized resource utilization.

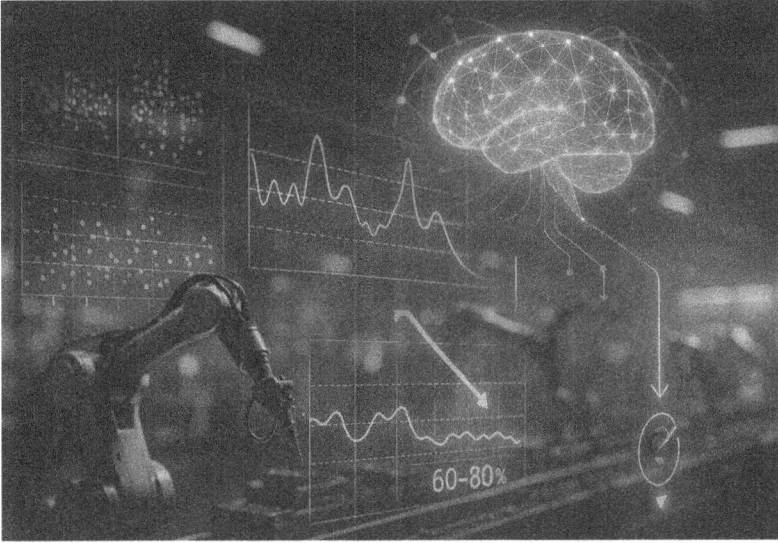

Real-time monitoring powered by AI delivers immediate insights into production quality status, enabling instant responses to process deviations. Advanced analytics systems process thousands of data points per second, identifying subtle trends indicating potential quality issues hours before traditional methods. Predictive analytics forecast equipment maintenance requirements, process optimization opportunities, and quality risk factors using historical data patterns. Benefits include reduced downtime through predictive maintenance scheduling, lower scrap rates via early intervention, and improved overall equipment effectiveness through optimized operating parameters. Implementation requires robust data infrastructure, skilled analysts, and management commitment to data-driven decision making. Organizations achieving success report quality improvements of 40-70% within first year of AI-enhanced SPC implementation.

Chapter 3: Quality Control Automation

Automated visual inspection systems comprise four integrated components working synergistically to achieve superior quality control. High-resolution cameras capture detailed images essential for accurate defect detection, often operating at resolutions exceeding traditional photography standards. Specialized lighting setups eliminate shadows, enhance contrast, and reveal surface features invisible under normal illumination conditions. Advanced image processing software analyzes captured images using sophisticated algorithms that identify defects, measure dimensions, and classify quality status. Robotics and conveyor integration enables seamless product handling throughout inspection processes, maintaining production flow while ensuring comprehensive quality assessment. System architecture must accommodate varying product sizes, shapes, and materials while maintaining consistent inspection standards across all variants.

Automated inspection systems deliver three significant advantages over manual methods. Consistency eliminates human variability factors including fatigue, distraction, and subjective interpretation differences, ensuring identical inspection standards throughout production shifts. Speed advantages become apparent when systems inspect products at rates impossible for human operators, often exceeding 10x manual inspection speeds while maintaining superior accuracy. Enhanced data logging capabilities provide detailed traceability records for quality analysis, regulatory compliance, and continuous improvement initiatives. Digital documentation enables trend analysis, defect correlation studies, and process optimization insights previously unavailable with manual inspection methods. These advantages combine to create inspection capabilities surpassing human performance while reducing long-term operational costs through improved efficiency and reduced quality escapes.

Successful automated inspection deployment requires careful consideration of multiple factors affecting system performance and longevity. Industry applications span automotive, electronics, and food processing sectors, each requiring specialized adaptation for unique product characteristics and quality requirements. Environmental conditions significantly impact system performance, necessitating protection from temperature extremes, humidity variations, dust contamination, and vibration exposure. Scalability considerations ensure systems accommodate future production growth while maintaining inspection quality standards. Maintenance requirements must balance system availability with preventive care needs, requiring trained technicians and spare parts inventory management. Proper deployment planning addresses these factors comprehensively, ensuring sustainable operation and maximum return on investment while meeting evolving quality requirements.

Generative models represent advanced machine learning techniques capable of creating realistic synthetic data samples. These models learn underlying data distributions from training examples, enabling generation of new samples exhibiting similar statistical properties. Generative Adversarial Networks use competitive training between generator and discriminator networks, iteratively improving sample quality through adversarial learning processes. Variational Autoencoders encode data into compressed latent representations before probabilistically reconstructing samples, enabling controlled variation generation. Both architectures excel at anomaly generation by modeling normal data patterns then creating synthetic variations representing potential defects. Understanding these foundational concepts enables effective application of generative models for quality control enhancement through synthetic data augmentation.

Synthetic anomaly generation addresses critical challenges in training robust defect detection systems. Real-world defect data often exhibits severe class imbalance, with normal products vastly outnumbering defective samples needed for comprehensive model training. Synthetic anomaly creation augments limited defect datasets, providing diverse examples representing potential failure modes rarely observed in production. Improved model generalization results from exposure to broader defect variation ranges during training, enabling better performance on previously unseen defect types. Enhanced rare defect detection capabilities emerge when models train on synthetic examples representing low-frequency but high-impact quality issues. This approach transforms quality control from reactive defect catching to proactive defect prevention through comprehensive model preparation.

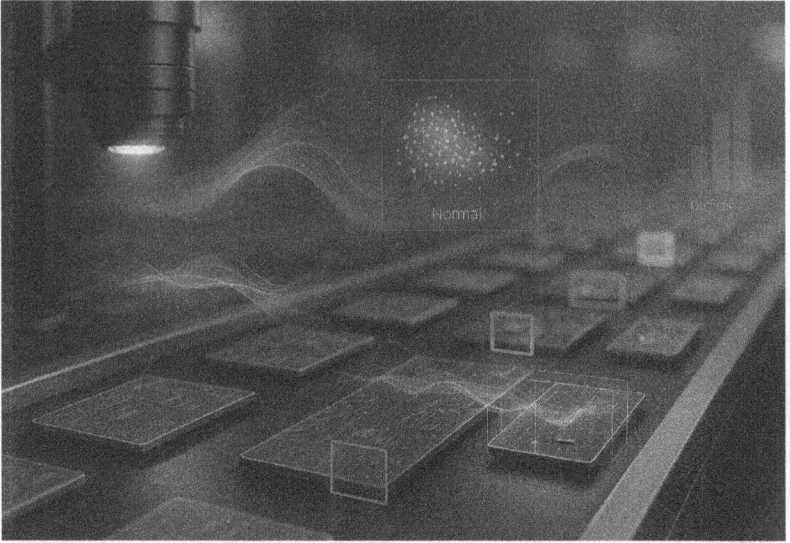

Synthetic data integration delivers measurable improvements in defect detection system performance across multiple key metrics. Enhanced model accuracy stems from balanced training datasets containing sufficient defect examples for robust learning, often improving detection rates by 20-40% compared to models trained solely on real data. Increased robustness emerges when models encounter diverse synthetic anomalies during training, preparing them for real-world variation ranges exceeding historical experience. Reduced false negative rates represent perhaps the most valuable improvement, as missed defects typically cost significantly more than false alarms through warranty claims, customer dissatisfaction, and brand reputation damage. Implementation requires careful validation ensuring synthetic data maintains realistic characteristics while providing meaningful learning value for detection algorithms.

Chapter 3: Quality Control Automation

In this chapter, we have explored four revolutionary technologies transforming manufacturing quality control. Computer vision enables microscopic defect detection at production speeds impossible with manual inspection, while AI-enhanced statistical process control predicts quality issues before they occur. Automated inspection systems provide consistent, traceable quality assessment eliminating human variability, and generative models create synthetic training data addressing real-world dataset limitations. These technologies represent the foundation of Industry 4.0 quality management, enabling predictive rather than reactive quality control approaches. Successful implementation requires strategic planning, skilled personnel, and commitment to continuous improvement. Organizations embracing these advanced techniques achieve superior quality outcomes while reducing operational costs and enhancing customer satisfaction.

Chapter 4: Process Optimization

Process optimization in manufacturing encompasses several proven methodologies, each offering unique advantages. Statistical Process Control forms the foundation by using data analysis to monitor and maintain consistent quality throughout production cycles. This approach helps identify variations before they become quality issues. Lean Manufacturing focuses on waste elimination and process flow improvement, maximizing value while minimizing resource consumption. Six Sigma methodology takes a disciplined, data-driven approach to reduce process variation and defects, targeting near-perfect quality levels. Finally, AI-driven approaches represent the cutting edge, utilizing machine learning and automation to optimize processes and enhance decision-making capabilities. These methodologies often work synergistically, with modern manufacturers combining traditional approaches with advanced AI techniques for maximum effectiveness.

Modern manufacturing faces both significant opportunities and complex challenges in optimization efforts. The benefits are substantial: optimized manufacturing processes deliver measurable cost savings, increased production throughput, and consistent product quality across all production batches. However, challenges persist. Manufacturing systems involve intricate interactions between multiple components, requiring careful management to maintain operational stability and efficiency. Raw material variability poses ongoing challenges to process consistency, directly impacting both product quality and manufacturing efficiency. Additionally, integrating new technologies with existing legacy systems presents implementation hurdles, though this integration is essential for modernization and improved performance. Understanding these trade-offs helps manufacturers make informed decisions about optimization investments and implementation strategies.

Artificial Intelligence integration represents a paradigm shift in process optimization capabilities. Predictive analytics enables manufacturers to forecast trends and anticipate process outcomes, supporting proactive planning and decision-making rather than reactive responses. Real-time monitoring systems powered by AI provide continuous oversight of manufacturing processes, instantly detecting deviations and enabling immediate corrective actions. This capability significantly reduces downtime and quality issues. Adaptive process controls represent perhaps the most sophisticated application, where AI systems dynamically adjust manufacturing processes to optimize performance continuously. These systems learn from ongoing operations, improving decision accuracy over time. The integration of these AI capabilities creates intelligent manufacturing environments that can self-optimize, adapt to changing conditions, and maintain peak performance with minimal human intervention.

Reinforcement learning operates on fundamental principles that make it particularly well-suited for manufacturing optimization. The core concept involves an agent that continuously interacts with its environment - in this case, the manufacturing system - to gather information essential for learning and decision-making. The reward feedback mechanism provides crucial guidance, offering positive or negative reinforcement based on the outcomes of actions taken. This creates a learning loop where the system improves through experience. The ultimate goal is optimal action learning, where the agent develops the ability to choose actions that maximize cumulative rewards over extended periods. This approach mimics natural learning processes, allowing manufacturing systems to adapt and improve continuously. The beauty of reinforcement learning lies in its ability to handle complex, dynamic environments without requiring explicit programming for every possible scenario.

Reinforcement learning applications in automated production lines demonstrate remarkable versatility and effectiveness. Optimizing scheduling represents a primary application, where algorithms dynamically adapt to changing production demands, minimizing delays and maximizing throughput efficiency. The system learns to balance multiple competing priorities in real-time. Resource allocation efficiency is another critical area, where intelligent algorithms distribute manufacturing resources - including materials, equipment, and personnel - to enhance overall productivity while reducing waste. This optimization occurs continuously as conditions change. Robotic control applications showcase perhaps the most visible benefits, where reinforcement learning enables robots to develop and refine optimized control strategies that significantly boost production line performance. These robots learn from experience, becoming more efficient and precise over time, ultimately outperforming traditional programmed approaches.

Case studies demonstrate tangible performance improvements from reinforcement learning implementation. Improved throughput represents one of the most significant benefits, with reinforcement learning systems optimizing production processes and resource allocation to achieve measurable increases in manufacturing output. These improvements often exceed 15-20% in well-implemented systems. Reduced energy consumption addresses both cost and sustainability concerns, as intelligent control systems powered by reinforcement learning algorithms can significantly lower energy usage while maintaining or improving production levels. Enhanced flexibility provides perhaps the greatest competitive advantage, allowing adaptive manufacturing systems to respond quickly and effectively to changing production demands, market conditions, and customer requirements. This adaptability enables manufacturers to remain competitive in rapidly changing markets while maintaining operational efficiency and quality standards throughout the adjustment process.

Process mining techniques provide unprecedented visibility into actual manufacturing operations. Event log analysis forms the foundation by using detailed operational data to reconstruct real process flows, revealing how work actually moves through the system versus how it was designed to flow. This analysis often uncovers surprising inefficiencies and improvement opportunities. Bottleneck detection capabilities systematically identify constraint points within processes, enabling targeted performance improvements where they will have maximum impact. These bottlenecks often shift as improvements are made, requiring continuous monitoring. Deviation identification helps maintain process compliance by detecting variations from standard operating procedures, ensuring both quality consistency and regulatory compliance. Finally, these insights provide a factual optimization basis, offering concrete data-driven evidence for process improvement initiatives. This evidence-based approach ensures that optimization efforts focus on real issues rather than assumed problems.

AI-enhanced process discovery capabilities significantly expand traditional process mining effectiveness. Automated pattern recognition eliminates manual analysis bottlenecks by enabling AI systems to identify patterns in complex manufacturing data automatically. This automation allows for faster and more comprehensive process discovery across larger datasets than human analysis could achieve. Anomaly detection systems reveal irregularities and potential inefficiencies early in their development, preventing small issues from becoming major problems. These systems can detect subtle patterns that human observers might miss. Predictive analytics takes process mining beyond historical analysis by forecasting future process outcomes and supporting proactive decision-making. This forward-looking capability enables manufacturers to anticipate and prevent problems rather than simply responding to them after they occur. The combination of these AI capabilities creates a powerful process intelligence platform.

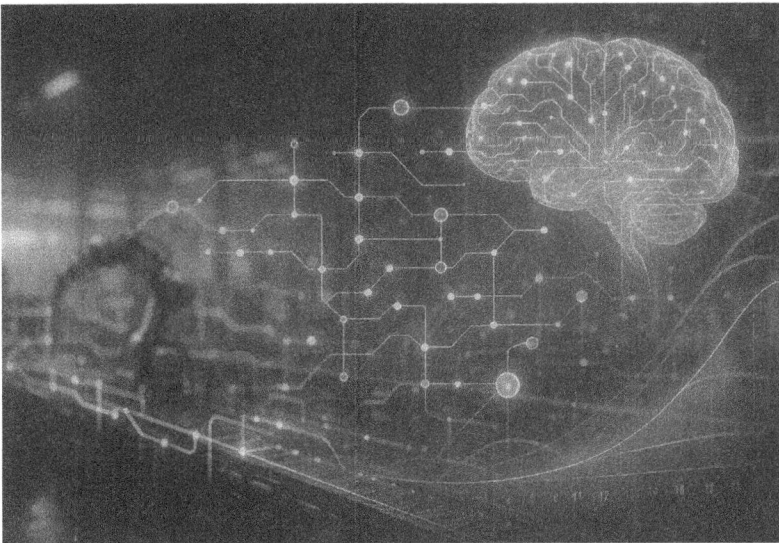

Intelligent analysis enables comprehensive workflow optimization that delivers measurable operational improvements. Data-driven workflow redesign uses insights gained from process mining to restructure manufacturing workflows for enhanced operational efficiency. This redesign is based on actual performance data rather than theoretical models, ensuring improvements are practical and achievable. Minimizing delays becomes possible through optimized workflows that reduce production bottlenecks and waiting times, leading to faster turnaround times and improved delivery performance. Customer satisfaction typically improves as a direct result. Reducing waste and improving resource utilization represents a dual benefit, where workflow redesign helps minimize material waste while simultaneously improving the effective use of available resources including equipment, materials, and human resources. This optimization often reveals hidden capacity within existing operations, delaying or eliminating the need for additional capital investments.

Real-time parameter adjustment strategies represent sophisticated manufacturing control systems that respond dynamically to changing conditions. Adaptive strategy overview encompasses the use of real-time operational data to adjust critical manufacturing parameters continuously, ensuring enhanced system performance under varying conditions. These systems never stop learning and adjusting. Sensor data utilization provides the foundation through continuous monitoring of critical conditions including temperature, pressure, speed, and other process variables. This constant data stream provides the feedback necessary for intelligent decision-making. Feedback loop mechanisms analyze incoming sensor data to make continuous fine-tuning adjustments, ensuring optimal operating conditions are maintained regardless of external variations. These loops operate at speeds impossible for human operators, making thousands of micro-adjustments per minute. The result is unprecedented process stability and optimization that adapts automatically to changing conditions without human intervention.

Machine learning applications for adaptive control represent the cutting edge of manufacturing automation and optimization. Predictive system behavior capabilities enable machine learning models to analyze complex operational data and accurately forecast system behavior, enabling timely interventions before problems develop. This predictive capability transforms maintenance from reactive to proactive. Parameter adjustment recommendations provide real-time guidance by suggesting specific parameter changes to optimize performance and reduce errors as conditions change. These recommendations are based on continuous learning from system performance. Proactive defect prevention uses predictive insights to prevent quality issues before they occur, significantly improving system reliability and product consistency. Efficiency maximization ensures that optimized control parameters continuously maximize system efficiency while reducing both waste and energy consumption. This holistic approach to optimization addresses multiple objectives simultaneously, delivering comprehensive performance improvements across all operational dimensions.

Real-world examples demonstrate the tangible benefits of improved efficiency and quality through adaptive parameter control systems. Waste reduction achievements show how adaptive parameter control helps industries significantly reduce material waste by optimizing resource utilization and minimizing defects during production processes. These reductions often exceed 20-30% of previous waste levels. Product consistency improvements result from enhanced control systems that ensure reliable quality across all manufacturing batches, eliminating the variation that previously caused quality issues and customer complaints. This consistency builds brand reputation and reduces costs associated with rework and returns. Responsive production capabilities represent perhaps the most valuable benefit, where adaptive controls increase manufacturing responsiveness to changing market demands, allowing flexible and efficient production adjustments without sacrificing quality or efficiency. This responsiveness provides significant competitive advantages in rapidly changing markets.

 Multi-objective optimization addresses the complex reality of manufacturing where multiple competing goals must be balanced simultaneously. Trade-offs in manufacturing require careful consideration as manufacturers must balance speed, cost, quality, and sustainability objectives to meet diverse production goals effectively. These objectives often conflict, requiring sophisticated optimization approaches. For example, increasing speed might compromise quality, or reducing costs might impact sustainability initiatives. Multi-objective optimization frameworks provide systematic approaches to find the best possible compromise among these competing production factors, ensuring better overall outcomes than focusing on any single objective. These frameworks use advanced algorithms to explore the solution space and identify optimal trade-off points. The goal is not to maximize any single metric, but to find the optimal balance that delivers the best overall performance across all critical business objectives.

 AI-driven multi-objective algorithms provide sophisticated tools for handling complex optimization challenges in manufacturing environments. Genetic algorithms simulate natural selection processes to find optimal solutions across multiple objectives simultaneously, particularly effective for complex manufacturing tasks where traditional optimization methods fall short. These algorithms evolve solutions over multiple generations, gradually improving performance. Neural networks excel at modeling complex patterns and relationships, enabling optimization of manufacturing parameters while considering various criteria simultaneously. Their ability to handle non-linear relationships makes them particularly valuable for complex manufacturing systems. Multi-criteria optimization capabilities allow these AI methods to evaluate vast solution spaces efficiently, identifying parameter sets that meet multiple manufacturing objectives simultaneously. This comprehensive approach ensures that optimization efforts consider all critical factors rather than sub-optimizing individual components. The result is more balanced, sustainable, and effective manufacturing operations.

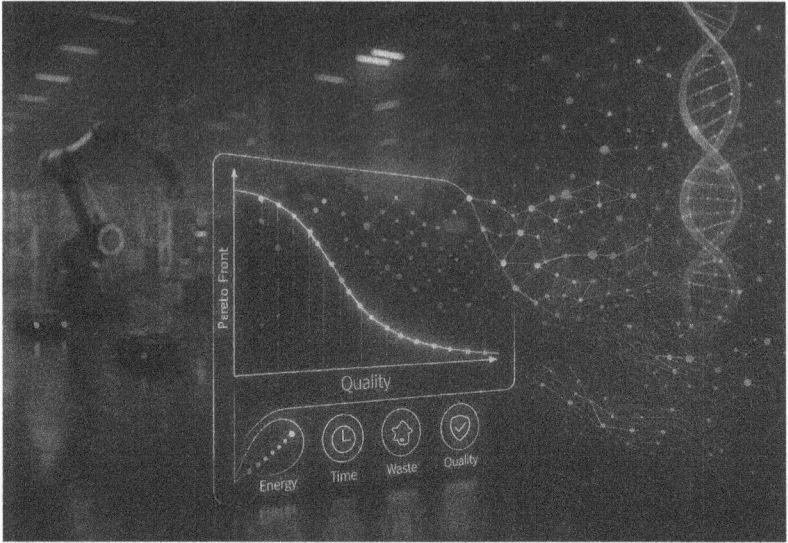

Practical manufacturing scenarios demonstrate the real-world application and benefits of multi-objective optimization approaches. Scheduling optimization applications show how optimizing production schedules reduces lead times while ensuring quality standards are maintained effectively, balancing speed and quality objectives that traditionally compete. This optimization considers resource constraints, quality requirements, and delivery commitments simultaneously. Energy management represents another critical application area, where balancing energy consumption with production output helps enhance both operational efficiency and environmental sustainability in manufacturing operations. These systems optimize energy usage patterns while maintaining production targets, often reducing energy costs by 15-25%. The integration of these optimization approaches creates manufacturing systems that perform better across multiple dimensions simultaneously, rather than excelling in one area at the expense of others. This balanced approach supports long-term business sustainability and competitive advantage.

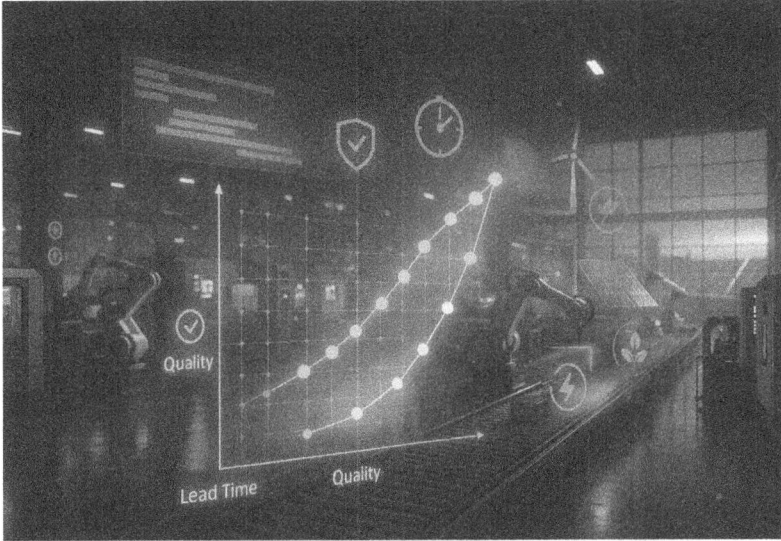

Simulation plays a crucial role in modern process optimization, providing capabilities that would be impossible or impractical with real-world testing alone. Safe testing environments allow manufacturers to test process changes and AI strategies without risk to actual production operations, equipment, or personnel. This safety enables more aggressive experimentation and innovation. Cost-effective innovation becomes possible as simulations reduce the expenses associated with process optimization by eliminating the need for expensive real-world trials and potential production disruptions. Organizations can test dozens of scenarios at a fraction of the cost of physical trials. Accelerated innovation cycles result from simulation capabilities that allow rapid experimentation and refinement of optimization strategies. What might take months to test in the real world can be evaluated in days or weeks through simulation. This acceleration enables faster deployment of improvements and more rapid response to changing market conditions and customer requirements.

Training AI models with simulated data addresses several critical challenges in manufacturing AI implementation. Addressing data scarcity problems, simulated data provides abundant training examples in situations where real operational data is limited, unavailable, or insufficient for effective AI model training. This is particularly valuable for new processes or rare operating conditions. Enhancing data diversity through simulation introduces varied scenarios and edge cases that improve AI model generalization to real-world situations, making the models more robust and reliable. Controlled training environments provide safe spaces for testing and optimizing AI models before real-world deployment, ensuring reliability and performance before implementation. The ability to generate unlimited training data under controlled conditions enables more thorough model development and validation. This comprehensive training approach results in more reliable, robust, and effective AI systems that perform better when deployed in actual manufacturing environments.

Understanding both benefits and limitations of simulation-augmented approaches ensures realistic expectations and effective implementation strategies. Training improvement benefits show how simulations enhance training effectiveness by providing realistic, safe practice environments for both human operators and AI systems. These environments enable skill development without production risks. Cost reduction advantages demonstrate how simulation use reduces operational costs by minimizing the need for physical resources during training and development phases. However, model inaccuracies present real limitations, as simulation models can contain errors that affect the reliability of outcomes and decisions. Regular validation against real-world performance is essential. The reality gap represents perhaps the most significant challenge, where differences between simulated and actual conditions require careful management to ensure effectiveness. Successful implementation requires ongoing calibration and validation to maintain simulation accuracy and relevance to real operating conditions.

In this chapter, we've explored four transformative approaches to modern manufacturing optimization. Reinforcement learning enables machines to learn optimal manufacturing decisions through continuous trial and feedback mechanisms, creating adaptive systems that improve over time. Process mining analyzes actual manufacturing workflows to identify hidden inefficiencies and improvement opportunities, providing data-driven insights for optimization initiatives. Adaptive controls adjust manufacturing parameters in real-time for enhanced quality and efficiency, responding instantly to changing conditions. Simulation-augmented training enhances both operator skills and AI system capabilities, preparing them for complex manufacturing challenges while reducing risk and cost. Together, these advanced techniques represent the future of manufacturing optimization, offering unprecedented opportunities for efficiency, quality, and competitive advantage. Implementation of these approaches requires careful planning, but the potential benefits make them essential considerations for modern manufacturing operations seeking sustainable competitive advantage.

Chapter 5: Energy Efficiency and Sustainability

Modern industries face unprecedented energy challenges that threaten both profitability and sustainability. Rising energy demands are driven by industrial growth and increased production requirements, placing enormous strain on existing infrastructure. Simultaneously, escalating energy costs are eroding profit margins and forcing companies to seek innovative cost-reduction strategies. The environmental impact of industrial energy consumption has become a critical concern, contributing significantly to greenhouse gas emissions and environmental degradation. These converging pressures create an urgent need for innovative solutions that can optimize energy use while maintaining operational effectiveness and reducing environmental footprints across all industrial sectors.

Chapter 5: Energy Efficiency and Sustainability

The transition to sustainable energy systems has become essential for addressing global climate challenges. Mitigating climate change requires sustainable energy systems that significantly reduce greenhouse gas emissions through cleaner energy production and consumption patterns. Renewable energy integration is vital for long-term sustainability, incorporating solar, wind, and other clean energy sources into existing infrastructure. However, this integration presents unique challenges in managing variable energy outputs. Reducing waste and improving efficiency ensures both economic viability and environmental sustainability. These interconnected goals require sophisticated management systems that can balance multiple objectives while maintaining reliable energy supply and supporting economic growth.

Artificial intelligence is transforming energy management through three primary applications. AI monitoring systems continuously track energy usage patterns, providing real-time insights that enable immediate optimization of consumption patterns and identification of inefficiencies. Machine learning algorithms optimize energy distribution across complex networks, reducing waste through intelligent routing and load management. Predictive energy management capabilities allow AI systems to forecast future demand trends, enabling proactive planning and resource allocation. These AI applications work synergistically to create intelligent energy management systems that adapt to changing conditions, optimize performance automatically, and support sustainable energy practices while maintaining operational reliability and cost-effectiveness.

AI predicts energy consumption patterns through sophisticated data analysis and forecasting methodologies. Historical data analysis forms the foundation, with AI systems examining years of past energy usage to identify consumption trends, seasonal variations, and cyclical patterns. Weather forecast integration enhances prediction accuracy by incorporating meteorological data that significantly influences energy demand, particularly for heating and cooling systems. User behavior analysis provides insights into consumption patterns based on human activities, occupancy rates, and operational schedules. These combined data sources enable proactive energy management, allowing utilities and facility managers to optimize resource allocation, prevent shortages, and improve overall system efficiency through data-driven decision-making.

Accurate demand forecasting delivers substantial benefits for utility operations and system reliability. Optimized resource allocation ensures utilities can efficiently distribute resources, preventing both waste and shortages while matching supply precisely to demand requirements. Cost reduction occurs through precise forecasting that minimizes unnecessary power generation, reduces maintenance expenses, and eliminates emergency procurement costs. Increased reliability and service quality result from accurate demand predictions that improve grid stability and prevent outages through better planning. Lower environmental impact is achieved by optimizing energy production levels, reducing emissions from excess generation, and supporting more efficient integration of renewable energy sources into the overall energy portfolio.

Real-world case studies demonstrate AI forecasting's practical benefits across multiple applications. Energy planning enhancement showcases how AI forecasting optimizes both production and distribution systems, enabling utilities to plan more efficiently and manage resources effectively across complex networks. Grid stability improvement illustrates AI algorithms' ability to enhance system reliability by predicting demand fluctuations and preventing potentially damaging outages through proactive management. Practical benefits and scalability demonstrate that real-world AI forecasting applications are not only effective but also scalable across different system sizes and geographic regions. These implementations provide tangible evidence of AI's transformative impact on energy systems, showing measurable improvements in efficiency, reliability, and cost-effectiveness across diverse operational environments.

Modern emissions monitoring relies on advanced technology and AI-powered analysis for accurate environmental tracking. Advanced sensor technology enables precise detection of various emission sources and real-time measurement of pollutant levels across multiple locations and emission types. These sophisticated sensors can identify specific compounds, measure concentrations accurately, and provide continuous monitoring capabilities. AI-powered data analysis processes the vast amounts of sensor data to provide comprehensive insights for emissions monitoring, identifying patterns, anomalies, and trends that would be impossible to detect through manual analysis. This combination of advanced sensing and intelligent analysis creates robust monitoring systems that support environmental compliance, regulatory reporting, and proactive environmental management.

Machine learning models provide sophisticated emissions prediction capabilities through comprehensive data analysis. Historical data analysis allows these models to examine past emission patterns, identifying trends and correlations that inform future predictions with increasing accuracy over time. Predictive forecasting capabilities enable organizations to anticipate emission levels before they occur, allowing for proactive management and intervention strategies. Environmental impact mitigation becomes possible through these forecasting capabilities, as organizations can implement preventive strategies to reduce emissions before they reach harmful levels. These predictive models support both regulatory compliance and environmental stewardship by enabling data-driven decision-making and proactive environmental management that goes beyond reactive compliance measures.

Integrating emissions data for regulatory compliance requires sophisticated data management and reporting systems. AI-driven data aggregation efficiently collects and consolidates emissions information from multiple sources, sensors, and monitoring points into comprehensive databases that provide complete visibility into organizational emissions. This automated aggregation eliminates manual data collection errors and ensures consistent, accurate reporting. Regulatory compliance reporting ensures that emissions data meets environmental standards and regulatory requirements through automated reporting systems that generate compliant documentation. These AI systems maintain audit trails, ensure data integrity, and provide the documentation necessary for regulatory submissions, environmental impact assessments, and sustainability reporting that meets increasingly stringent environmental regulations.

Chapter 5: Energy Efficiency and Sustainability

Grid load management faces significant challenges that require sophisticated solutions for modern energy systems. Variable demand fluctuations create unpredictable changes in consumer energy requirements, complicating load management and supply planning processes that must balance supply and demand continuously. Renewable energy integration introduces additional variability through wind and solar sources that fluctuate based on weather conditions, creating challenges in maintaining consistent grid stability. Balancing supply and demand becomes critical to prevent outages and ensure reliable service delivery. These interconnected challenges require intelligent management systems that can respond rapidly to changing conditions while maintaining system stability and preventing costly disruptions to energy service.

AI strategies enable real-time load balancing through sophisticated analysis and optimization techniques. Real-time data analysis involves AI algorithms continuously monitoring grid conditions to detect load variations and predict demand changes instantly, enabling immediate response to system conditions. Optimized load distribution uses AI to manage power flow efficiently, preventing overloads and balancing electrical loads across the entire grid infrastructure. Enhanced grid responsiveness results from AI systems that adapt quickly to changing conditions and demand fluctuations, automatically adjusting distribution patterns to maintain stability. These AI-driven strategies work together to create intelligent grid management systems that respond faster and more accurately than traditional methods while maintaining reliability.

Chapter 5: Energy Efficiency and Sustainability

AI's impact on grid stability and renewable integration creates multiple benefits for modern energy systems. Smarter load management optimizes electricity distribution by effectively managing loads to prevent overloads, outages, and system instabilities while maintaining consistent service quality. Renewable energy integration benefits from AI systems that stabilize grids despite fluctuating power inputs from wind and solar sources, enabling higher renewable energy adoption rates. Grid stability preservation ensures continuous, reliable power supply by reducing interruptions and maintaining the critical balance between supply and demand. These AI-driven improvements support the transition to cleaner energy sources while maintaining the reliability that consumers and businesses require for daily operations.

Data analytics enables comprehensive identification of operational inefficiencies through intelligent analysis systems. AI data analysis examines vast amounts of operational information to identify inefficiencies and waste patterns that would be impossible to detect manually. Pattern detection capabilities allow AI systems to identify recurring waste patterns, helping organizations focus improvement efforts on the most critical areas for maximum impact. Targeted interventions become possible when inefficiencies are precisely identified, enabling organizations to implement specific solutions that enhance operational performance and productivity. This systematic approach to inefficiency identification transforms operational management from reactive problem-solving to proactive optimization that continuously improves system performance and resource utilization.

Process optimization strategies focus on reducing energy waste through intelligent analysis and system improvements. Machine learning in manufacturing analyzes production data to identify operational inefficiencies and recommend specific energy-saving process modifications that can be implemented without disrupting production schedules. Energy system optimization uses AI algorithms to minimize unnecessary energy consumption while maintaining or enhancing operational efficiency across all system components. These optimization strategies work at multiple levels, from individual equipment optimization to facility-wide energy management, creating comprehensive efficiency improvements. The result is reduced energy costs, lower environmental impact, and improved operational sustainability that supports both economic and environmental objectives.

Industry examples demonstrate practical waste minimization through AI implementation across various sectors. AI systems reduce energy waste by monitoring and minimizing consumption patterns, cutting operational costs while maintaining productivity levels and service quality. Productivity increases result from AI-driven automation that enhances manufacturing efficiency by optimizing processes and reducing costly downtime through predictive maintenance and intelligent scheduling. Sustainability enhancement occurs as AI supports environmentally responsible practices by reducing waste and promoting resource-efficient operations that minimize environmental impact. These real-world applications show that AI-driven waste minimization delivers measurable benefits across multiple performance dimensions, supporting both operational excellence and environmental stewardship.

In conclusion, AI-driven innovations are transforming energy efficiency and sustainability across multiple critical areas. Improved forecasting capabilities enable better resource allocation and waste reduction through accurate demand prediction and intelligent planning systems. Emissions tracking provides precise monitoring that supports climate goals and ensures regulatory compliance through comprehensive data collection and analysis. Smart grid management optimizes system stability while facilitating renewable energy integration that supports clean energy transitions. Waste reduction strategies minimize energy consumption and improve overall sustainability performance across industrial operations. These interconnected AI applications create comprehensive energy management systems that address current challenges while supporting future sustainability goals and operational excellence.

Chapter 6: Demand Forecasting

Demand forecasting serves as the foundation of efficient supply chain management across four key dimensions. Production planning becomes significantly more effective when organizations can accurately predict customer needs, enabling better resource allocation and capacity utilization. Inventory management benefits tremendously from precise forecasts, as they help maintain optimal stock levels while reducing both excess inventory and costly stockouts. Waste reduction occurs naturally when supply aligns with actual demand, optimizing resource utilization and minimizing unnecessary costs. Most importantly, strategic decision making improves when leaders have reliable demand insights, supporting long-term supply chain profitability and competitive positioning in dynamic markets.

Despite its importance, accurate demand prediction faces significant challenges that organizations must address systematically. Data volatility presents ongoing difficulties as unpredictable changes and fluctuations in historical patterns complicate forecasting models. Seasonality effects require sophisticated modeling to account for recurring demand changes throughout different time periods. Market shifts can suddenly disrupt established demand trends, requiring forecasting systems to be both adaptable and responsive to changing conditions. External disruptions, such as economic downturns or environmental events, impact demand unpredictably and necessitate real-time data integration capabilities. Successfully navigating these challenges requires advanced forecasting techniques and flexible analytical approaches.

Chapter 6: Demand Forecasting

Classical statistical methods form the backbone of traditional forecasting approaches, with ARIMA and exponential smoothing serving as foundational techniques for modeling time series data. These methods excel at capturing underlying trends and seasonal patterns, making them valuable for businesses with relatively stable demand patterns. Their primary strength lies in interpretability, allowing analysts to understand and explain the forecasting logic to stakeholders. However, these classical approaches have notable limitations when dealing with complex or nonlinear patterns that extend beyond basic trends and seasonality. While they provide excellent starting points for forecasting initiatives, organizations with complex demand patterns may need to supplement these methods with more advanced techniques.

Machine learning approaches revolutionize time series forecasting by capturing complex nonlinear relationships and interactions within data that classical methods often miss. These advanced techniques excel at identifying subtle patterns and dependencies that traditional statistical methods cannot detect. Random forests and gradient boosting represent powerful ensemble methods that improve prediction accuracy by combining multiple models, reducing the risk of overfitting while enhancing overall performance. Neural networks offer sophisticated pattern recognition capabilities, enabling organizations to model highly complex time series relationships. These machine learning approaches are particularly valuable for businesses with intricate demand patterns, multiple influencing factors, or rapidly changing market conditions requiring advanced analytical capabilities.

Effective model evaluation and selection requires a comprehensive framework considering multiple critical factors. Accuracy metrics such as Mean Absolute Error, Root Mean Square Error, and Mean Absolute Percentage Error provide quantitative measures of forecasting performance, helping organizations compare different modeling approaches objectively. Model complexity considerations ensure that selected models neither overfit to historical data nor oversimplify complex relationships, striking the right balance for reliable predictions. Scalability and interpretability represent crucial practical considerations, as models must handle growing data volumes while remaining understandable for effective decision-making. The best forecasting approach balances these factors, providing accurate, scalable, and interpretable solutions that meet organizational needs and capabilities.

Chapter 6: Demand Forecasting

Market and economic indicators provide valuable external signals that significantly enhance forecasting accuracy when properly integrated into prediction models. Market trends analysis helps organizations identify emerging demand patterns and anticipate shifts in the economic environment before they fully materialize. Economic indicators offer measurable data points to assess overall economic health, providing context for predicting future market movements and consumer behavior. Consumer sentiment reflects public confidence levels and directly influences purchasing decisions, making it a crucial factor in demand forecasting. By incorporating these external signals, organizations can develop more robust forecasting models that account for broader market dynamics beyond historical sales data.

Integrating diverse external data sources creates comprehensive forecasting models that respond to real-world conditions affecting demand patterns. Weather conditions significantly influence demand variations across many industries, from retail clothing to food and beverage, requiring systematic integration of meteorological data into forecasting models. Social media trends provide real-time insights into consumer preferences and emerging behaviors, enabling organizations to anticipate demand changes before they appear in traditional sales data. Competitor actions and strategic moves impact market dynamics and customer choices, making competitive intelligence an essential component of advanced forecasting systems. Successfully combining these data sources creates more accurate and responsive forecasting capabilities.

Data quality, preprocessing, and normalization form the critical foundation for reliable forecasting systems, regardless of the analytical techniques employed. Ensuring data quality requires thorough cleaning and validation processes to identify and address inconsistencies, errors, and anomalies that could compromise forecast accuracy. Normalization techniques adjust data scales and formats to ensure comparability across different variables and time periods, improving algorithm performance and model stability. Addressing missing values and filtering noise are essential preprocessing steps that enhance data integrity and model reliability. Without proper data preparation, even the most sophisticated forecasting algorithms will produce unreliable results, making this foundational step absolutely crucial for successful demand prediction initiatives.

Artificial Intelligence transforms Sales and Operations Planning by introducing sophisticated automation and analytical capabilities across three key areas. Automated forecasting leverages AI to improve both accuracy and speed in predicting demand patterns, reducing manual effort while enhancing reliability of S&OP processes. Demand sensing capabilities enable AI systems to quickly analyze real-time data streams, providing better supply chain responsiveness and agility in dynamic market conditions. Supply planning optimization utilizes AI to reduce human bias and enable faster, more data-driven decision making, resulting in improved resource allocation and operational efficiency. These AI-powered capabilities collectively enhance the entire S&OP process, creating more responsive and effective planning systems.

AI-driven scenario planning and optimization provide organizations with powerful tools for managing uncertainty and improving decision-making capabilities. Scenario simulation enables AI systems to model various demand and supply scenarios efficiently, helping organizations prepare for multiple potential future outcomes and develop contingency plans. Inventory optimization leverages AI algorithms to determine optimal inventory levels that balance supply and demand requirements while reducing waste and minimizing stockout risks. Resource allocation benefits from AI's ability to optimize the distribution of materials and workforce under uncertain conditions, ensuring efficient utilization regardless of changing circumstances. These AI-driven capabilities enable organizations to plan more effectively and respond more rapidly to changing market conditions.

AI insights facilitate better alignment among cross-functional teams by providing transparent, data-driven information that supports collaborative decision-making. AI-powered dashboards deliver real-time data and analytics to all stakeholders, creating a common foundation for transparent decision-making across different departments and functions. Collaborative platforms enabled by AI facilitate seamless communication and coordination among sales, operations, and finance teams, breaking down traditional silos and improving organizational alignment. Enhanced decision transparency results from AI's ability to provide clear, objective insights that improve alignment and trust among cross-functional teams. By democratizing access to high-quality information and analytics, AI helps organizations create more cohesive and effective planning processes.

Identifying and assessing inventory risks requires systematic analysis of three primary risk categories that can significantly impact business operations. Demand variability creates ongoing challenges for inventory planning and stock management, as fluctuations in customer demand make it difficult to maintain optimal inventory levels. Supply disruptions represent critical risks as interruptions in the supply chain can delay inventory replenishment and increase the likelihood of costly stockouts. Obsolescence risk occurs when inventory becomes outdated or unsellable over time, particularly problematic for businesses dealing with technology products, fashion items, or perishable goods. Understanding these risk categories enables organizations to develop targeted mitigation strategies and build more resilient inventory management systems.

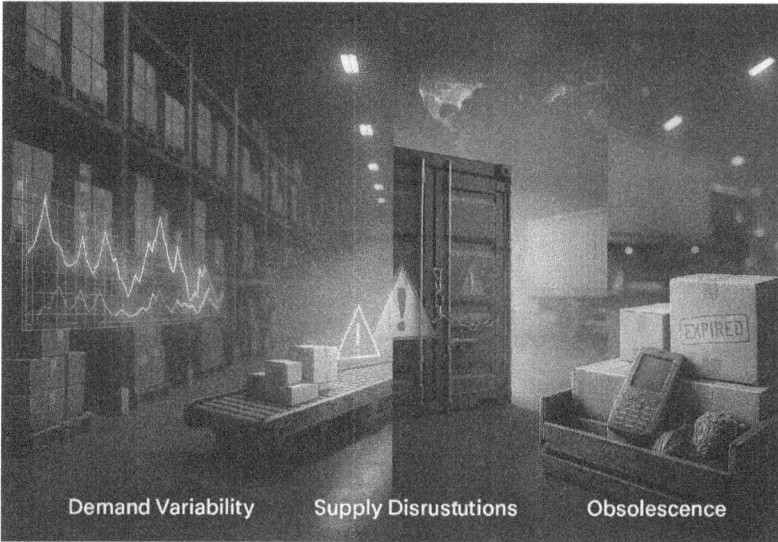

Demand Variability   Supply Disrustutions   Obsolescence

Effective risk mitigation requires implementing comprehensive strategies that address the various sources of inventory risk systematically. Safety stock calculations provide a quantitative approach to maintaining buffer inventory that protects against demand and supply variability, balancing service levels with inventory costs. Supplier diversification reduces dependency risks by establishing relationships with multiple suppliers for critical materials, ensuring continuity of supply even when individual suppliers face disruptions. Flexible replenishment strategies enable organizations to rapidly adjust inventory levels based on changing demand conditions, improving responsiveness while minimizing both excess inventory and stockout risks. These mitigation strategies work together to create more robust and resilient inventory management systems.

Continuous monitoring and improvement processes ensure that inventory management systems remain effective and adapt to changing business conditions over time. Performance tracking through continuous monitoring enables real-time assessment of inventory performance and operational metrics, providing immediate visibility into system effectiveness and identifying areas for improvement. Feedback loops provide essential data flows that enable organizations to adjust and optimize inventory policies dynamically based on actual performance results. Dynamic policy adjustment ensures that inventory management approaches evolve continuously to respond effectively to changing market conditions, customer requirements, and operational constraints. This continuous improvement approach helps organizations maintain optimal inventory performance while adapting to evolving business environments.

In conclusion, advanced approaches to demand forecasting and inventory risk management can transform supply chain performance. Advanced demand forecasting utilizing sophisticated statistical and machine learning models dramatically improves accuracy in predicting future product demand. Inventory risk management techniques effectively reduce both stockout situations and overstock problems through systematic risk assessment and mitigation. AI and external data integration enhance decision-making capabilities and organizational responsiveness to market changes. When combined, these techniques empower organizations to optimize supply chain performance for evolving markets, creating competitive advantages and improved operational efficiency in today's dynamic business environment.

Chapter 7: Procurement and Supplier Intelligence

Modern procurement has evolved far beyond traditional manual processes. Today's procurement landscape leverages digital workflow automation to eliminate repetitive tasks and reduce human error while accelerating decision-making cycles. Enhanced transparency mechanisms provide stakeholders with real-time visibility into procurement activities, fostering accountability and compliance. Real-time collaboration tools break down silos between procurement teams, suppliers, and internal stakeholders, enabling seamless communication and faster issue resolution. Finally, systematic cost and quality optimization approaches use data-driven insights to strengthen supplier relationships while achieving better financial outcomes and improved product quality.

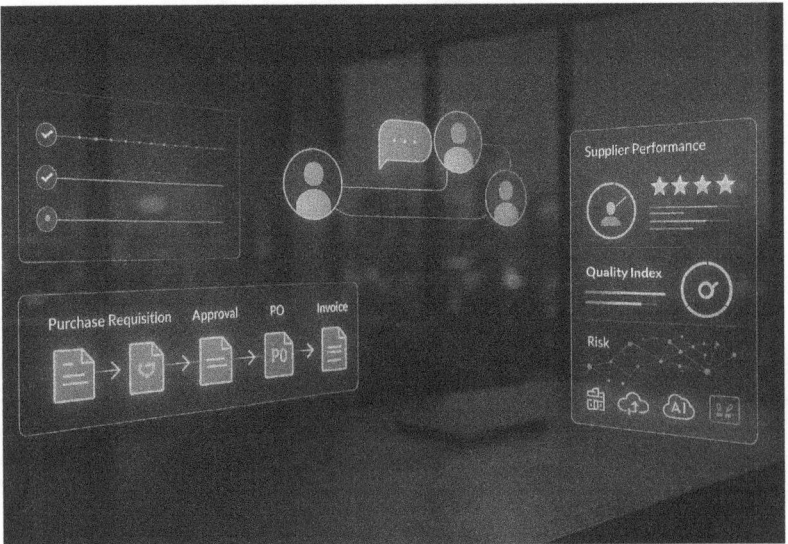

Supplier intelligence forms the backbone of strategic sourcing decisions. By tracking comprehensive supplier performance metrics, organizations gain deep insights into quality consistency, delivery reliability, and overall vendor capabilities. Financial health evaluation protects against supply chain disruptions by identifying potential risks before they materialize into operational challenges. Market position awareness enables procurement teams to understand supplier competitiveness, innovation capabilities, and long-term viability within their respective industries. These intelligence capabilities collectively support robust risk mitigation strategies, helping organizations build resilient supply chains that can adapt to market volatility and unexpected disruptions while maintaining operational continuity.

Technology integration represents a fundamental shift in procurement decision-making processes. Data analytics transforms raw procurement data into actionable insights, enabling evidence-based decisions that optimize both cost and quality outcomes. AI-powered automation handles routine procurement tasks such as purchase order processing, vendor communications, and compliance monitoring, freeing procurement professionals to focus on strategic activities. Cloud-based platforms revolutionize procurement operations by providing centralized access to supplier information, contract databases, and performance metrics while enabling seamless collaboration across distributed teams and improving overall transparency throughout the procurement lifecycle.

Effective risk assessment in supplier management relies on systematic data collection methods that capture comprehensive supplier information across multiple dimensions. These include financial stability indicators, operational performance metrics, compliance history, and market reputation factors. Quantitative scoring models provide objective frameworks for evaluating supplier risk levels, enabling consistent comparison across different vendors and categories. Scenario analysis techniques help procurement teams anticipate potential disruptions by modeling various risk scenarios and their potential impacts. This proactive approach enables organizations to develop contingency plans and mitigation strategies before problems occur, ensuring supply chain resilience and business continuity.

Machine learning models revolutionize vendor scoring by processing vast amounts of historical and real-time data to identify patterns and trends invisible to traditional analysis methods. These sophisticated algorithms analyze multiple data sources including financial performance, delivery history, quality metrics, and market conditions to generate comprehensive vendor insights. Predictive scoring capabilities enable procurement teams to anticipate supplier performance and identify potential issues before they impact operations. The objective nature of algorithmic vendor ranking eliminates human bias from supplier evaluation processes, ensuring fair and consistent assessments based on performance data, risk profiles, and strategic alignment with organizational objectives.

Predictive analytics delivers transformative benefits for procurement operations. Risk forecasting capabilities identify potential supply chain disruptions, financial instabilities, and compliance issues before they materialize, enabling proactive intervention strategies. Enhanced supplier selection processes leverage comprehensive performance and reliability data analysis to identify optimal vendor partnerships that align with organizational objectives. Cost reduction opportunities emerge through optimized purchasing decisions, improved inventory management, and strategic sourcing recommendations. Most importantly, predictive analytics enables truly proactive decision-making, empowering procurement teams to strengthen supply chain resilience, ensure business continuity, and maintain competitive advantage through data-driven strategic planning.

AI-powered contract review systems transform document analysis through automated scanning capabilities that rapidly identify key clauses, obligations, and critical terms within complex supplier agreements. Advanced algorithms excel at detecting inconsistencies, discrepancies, and potential compliance issues that might escape human review, significantly improving accuracy and reducing legal risks. The efficiency gains are substantial – automated contract review processes dramatically reduce manual workload while accelerating overall document processing timelines. This technological advancement allows legal and procurement teams to focus on strategic contract negotiation and relationship management rather than time-consuming document review tasks, ultimately improving both productivity and contract quality.

AI contract analysis systems excel at identifying and extracting critical contract components essential for effective supplier management. Pricing terms identification enables automated cost management and facilitates strategic negotiation planning by providing comprehensive visibility into financial obligations and payment structures. Delivery schedule detection ensures timely fulfillment expectations are clearly understood and tracked, supporting supply chain reliability and planning accuracy. Compliance requirements recognition maintains regulatory adherence by automatically flagging mandatory terms and monitoring ongoing obligations. Contract renewal date tracking prevents missed deadlines and supports proactive contract management, enabling timely renegotiations and ensuring continuous supplier relationships without service interruptions.

AI significantly reduces errors and improves compliance through comprehensive contract analysis automation. By automating review and analysis processes, AI systems eliminate human fatigue factors and provide consistent evaluation standards across all contract documents, dramatically reducing manual workload while accelerating processing timelines. The technology's ability to minimize human errors in contract processing stems from its consistent, objective evaluation methodology that doesn't suffer from oversight, distraction, or subjective interpretation issues. Most critically, AI enforces regulatory compliance by systematically verifying that contracts meet all applicable legal requirements and industry standards, ensuring that contractual obligations are clearly defined, properly documented, and fully enforceable.

Natural Language Processing technology excels at analyzing unstructured supplier communications, transforming emails, documents, and correspondence into actionable business intelligence. These sophisticated tools extract meaningful insights from large volumes of text-based communications that would be impossible to process manually. NLP systems identify key trends, issues, and patterns within communication data, supporting improved decision-making processes by highlighting recurring themes, emerging problems, and supplier sentiment indicators. Efficient categorization capabilities automatically organize supplier communications by topic, urgency, and required action, significantly improving operational efficiency and reducing response times while ensuring that critical supplier communications receive appropriate attention and timely resolution.

NLP insights significantly enhance supplier relationship management through sophisticated communication pattern analysis that reveals interaction frequency, response times, communication quality, and relationship health indicators. These insights enable procurement teams to better understand supplier interactions and optimize relationship management strategies. Advanced conflict resolution capabilities identify potential disputes and tension indicators early in the communication cycle, enabling faster intervention and resolution before issues escalate into serious relationship problems. The result is stronger, more resilient supplier partnerships built on data-driven understanding of communication patterns, preferences, and

81

relationship dynamics that foster trust, improve collaboration, and create sustainable competitive advantages through superior supplier relationship management.

Real-time sentiment analysis captures emotional cues and underlying attitudes instantly during supplier communications, providing procurement professionals with valuable insights into relationship health and negotiation dynamics. This technology enables more effective negotiation strategies by allowing teams to adjust tactics based on detected emotions, stress levels, and satisfaction indicators throughout the negotiation process. The benefits for procurement are substantial – sentiment insights help achieve better negotiation results by identifying optimal timing for proposals, recognizing when to address concerns, and understanding when relationships need attention. This emotional intelligence capability transforms supplier negotiations from purely transactional interactions into relationship-building opportunities that create long-term value for both parties.

Dynamic matching algorithms represent a paradigm shift in procurement supplier selection processes. These sophisticated systems utilize machine learning capabilities to analyze comprehensive supplier data sets, identifying optimal matches based on multiple criteria including performance history, capabilities, capacity, and risk profiles. Continuous supplier evaluation ensures that recommendations reflect real-time performance changes, market conditions, and evolving business requirements rather than static historical assessments. Advanced analytics provide the computational foundation for dynamic matching by processing real-time insights and predictive capabilities that enable procurement teams to make informed decisions quickly and adapt to changing market conditions, supplier availability, and evolving organizational needs.

Algorithmic supplier selection relies on comprehensive data sources and evaluation criteria to ensure optimal vendor matching. Historical supplier performance records provide reliability and quality assurance data that forms the foundation for selection decisions. Market data analysis incorporates competitive intelligence, pricing trends, and industry positioning information to assess supplier competitiveness and strategic value. Financial metrics evaluation examines supplier stability, viability, and long-term sustainability through key performance indicators and financial health assessments. Customer feedback integration captures service quality perspectives and satisfaction levels from multiple stakeholder viewpoints, ensuring that supplier selection decisions reflect comprehensive performance understanding rather than limited internal assessments.

Dynamic algorithms dramatically improve agility and responsiveness in supplier management by enabling real-time analysis and selection of optimal suppliers based on current conditions rather than historical assumptions. This technological capability allows procurement teams to quickly adapt to changing market conditions, supplier availability, and evolving business requirements. Enhanced market responsiveness emerges from technology-driven supplier recommendations that incorporate real-time market intelligence, pricing fluctuations, and capacity availability. The result is a more agile procurement function that can rapidly respond to opportunities, mitigate emerging risks, and optimize supplier relationships based on current rather than historical performance data, creating sustainable competitive advantages.

The technological revolution in procurement represents a fundamental transformation of how organizations manage supplier relationships and sourcing strategies. AI, machine learning, and NLP technologies are creating unprecedented opportunities for efficiency, accuracy, and strategic value creation in procurement operations. Advanced risk mitigation and sourcing optimization capabilities enable organizations to build resilient supply chains while achieving superior cost and quality outcomes. Enhanced contract and supplier management through these technologies creates competitive advantages by improving relationship quality, reducing operational risks, and enabling more strategic procurement decision-making that drives long-term organizational success and market positioning.

Chapter 8: AI-Driven Logistics and Warehousing

Machine learning algorithms analyze vast datasets including historical delivery patterns, real-time traffic conditions, and customer locations to determine optimal routes. This data-driven approach eliminates guesswork and human bias, resulting in consistently efficient path planning. The technology considers multiple variables simultaneously - distance, fuel consumption, vehicle capacity, and delivery windows - to create routes that minimize both time and cost. Dynamic adaptability allows the system to adjust routes in real-time based on changing conditions like traffic congestion, weather patterns, or urgent delivery requests, ensuring maximum efficiency throughout the day.

Real-time data integration transforms route optimization from static planning to dynamic, responsive navigation. Live traffic feeds enable AI systems to detect congestion patterns and automatically reroute vehicles to avoid delays, reducing delivery times by up to 25%. Weather data integration allows proactive planning for hazardous conditions like storms, snow, or flooding, protecting both cargo and drivers. The AI system continuously processes multiple data streams - GPS tracking, traffic sensors, weather satellites, and road condition reports - to make split-second routing decisions that human dispatchers couldn't match in speed or accuracy.

The financial impact of AI-driven route optimization is substantial and measurable. Fuel cost reduction typically ranges from 15-30% through shorter distances and optimized driving patterns. Labor costs decrease through reduced overtime and more efficient driver utilization. Delivery speed improvements of 20-40% enhance customer satisfaction and enable businesses to handle more deliveries with existing resources. These efficiency gains translate directly to competitive advantages - faster delivery times, lower operational costs, and improved service reliability. Customer satisfaction scores consistently improve due to accurate delivery windows and reduced delays.

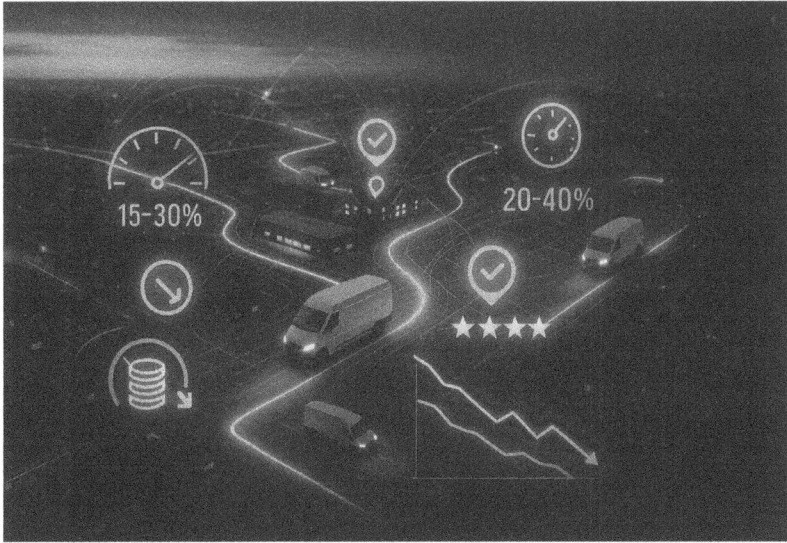

Automated Guided Vehicles represent the backbone of modern autonomous warehouses, operating independently using sophisticated navigation systems including laser guidance, magnetic strips, or vision-based technology. These vehicles transport materials between receiving, storage, picking, and shipping areas without human intervention, working continuously in 24/7 operations. AI-driven navigation enables precise positioning and obstacle avoidance, while task management systems optimize vehicle assignments based on priority, distance, and capacity. The integration reduces dependency on manual labor for material transport, eliminates human error in routing, and provides consistent, reliable movement of goods throughout warehouse facilities.

AI-enabled sorting and picking systems utilize computer vision, machine learning, and robotic manipulation to automate item identification and handling processes. Advanced cameras and sensors can distinguish between thousands of different products, reading barcodes, text, and visual characteristics with 99%+ accuracy. Machine learning algorithms continuously improve recognition capabilities, adapting to new products and packaging variations. The automation dramatically reduces manual labor requirements while increasing processing speed by 300-500%. Error rates drop significantly compared to human picking, reducing returns and customer complaints while improving overall order fulfillment accuracy and efficiency.

AI-powered safety monitoring systems use computer vision and sensor networks to continuously assess warehouse environments, detecting potential hazards before accidents occur. These systems monitor equipment performance, worker behavior, and environmental conditions, alerting supervisors to unsafe situations in real-time. Productivity gains result from continuous operations with minimal downtime - automated systems don't require breaks, shift changes, or training periods. The combination of enhanced safety protocols and 24/7 operational capability creates warehouses that are both safer for human workers and more productive overall, achieving efficiency levels impossible with traditional manual operations.

AI-powered sensors throughout warehouses provide real-time visibility into inventory levels, locations, and movement patterns. RFID tags, weight sensors, and computer vision systems automatically track items as they move through receiving, storage, and shipping processes. This continuous monitoring eliminates the need for time-consuming manual inventory counts while providing accurate, up-to-the-minute stock information. The system prevents stockouts by triggering automatic reorder points and reduces overstock situations by providing precise demand visibility. Inventory turnover improves through better demand prediction and optimal stock level maintenance, directly impacting cash flow and storage costs.

Predictive analytics transforms inventory management from reactive to proactive planning by analyzing historical sales data, seasonal patterns, market trends, and external factors to forecast future demand with remarkable accuracy. Machine learning algorithms identify complex patterns that human analysts might miss, considering variables like weather, economic indicators, and social trends. These accurate forecasts enable businesses to optimize inventory levels, reducing carrying costs while ensuring product availability. Strategic resource allocation becomes data-driven, allowing companies to position inventory closer to anticipated demand and adjust staffing levels based on predicted workload patterns.

AI warehouse layout analysis utilizes optimization algorithms to maximize space utilization and streamline material flow patterns. The system analyzes product velocity, storage requirements, and picking patterns to determine optimal placement strategies - high-velocity items positioned near shipping areas, complementary products stored together. Resource deployment optimization ensures equipment, personnel, and storage capacity are allocated efficiently based on real-time demand patterns. This intelligent space management can increase warehouse capacity by 20-30% without physical expansion while reducing travel time for picking operations, directly improving throughput and operational costs.

AI algorithms coordinate multiple robots simultaneously, creating a synchronized workforce that operates more efficiently than individual units. Task allocation systems consider robot capabilities, current locations, battery levels, and workload to assign jobs optimally. The scheduling minimizes robot idle time by ensuring continuous workflow and balanced workload distribution. Advanced algorithms predict maintenance needs and schedule tasks accordingly, preventing operational disruptions. This coordination maximizes throughput by eliminating bottlenecks and ensuring seamless handoffs between different robotic systems, creating a highly efficient automated ecosystem that operates with minimal human oversight.

Human-robot collaboration represents the optimal balance between automation efficiency and human adaptability. Collaborative systems enable safe operation in shared workspaces through advanced sensors and AI safety protocols that prevent collisions and injuries. Humans provide critical thinking, problem-solving abilities, and flexibility for handling exceptions, while robots contribute precision, consistency, and tireless operation for repetitive tasks. This combination leverages the strengths of both, creating work environments where productivity, safety, and job satisfaction all improve. Workers focus on higher-value activities while robots handle physically demanding or monotonous tasks.

Real-world implementations demonstrate measurable success across multiple metrics. Operational efficiency improvements typically range from 25-50% through optimized coordination and reduced downtime. Error reduction of 80-90% results from precise robotic operations and AI quality control systems. These improvements translate to tangible business outcomes: faster order fulfillment, reduced operational costs, improved customer satisfaction scores, and enhanced competitive positioning. Case studies consistently show return on investment within 12-18 months, with ongoing benefits including reduced labor costs, improved safety records, and increased processing capacity without facility expansion.

AI transformation of logistics and warehousing delivers comprehensive benefits across all operational areas. Route optimization reduces transportation costs by 15-30% while improving delivery speed and reliability. Automated handling and robotics streamline warehouse operations, increasing throughput while enhancing worker safety through reduced manual handling of heavy items. The integration of these AI technologies creates supply chains that are more responsive, efficient, and cost-effective. Organizations implementing comprehensive AI solutions typically see overall supply chain cost reductions of 20-25% combined with significant improvements in customer satisfaction and competitive positioning in their markets.

Chapter 9: AI-Assisted Design and Engineering

Generative design represents a fundamental shift in how we approach engineering problems. The process begins with clearly defining design goals, constraints, and parameters - essentially teaching the AI what success looks like. The AI then generates numerous design alternatives, often exploring solutions that human designers might never consider. This isn't random generation; it's intelligent exploration of the design space based on engineering principles and optimization algorithms. Finally, engineers evaluate these AI-generated alternatives, selecting the most promising designs for further development. This human-AI collaboration combines computational power with human expertise and judgment.

The benefits of generative design extend across multiple dimensions of engineering practice. Speed is perhaps the most obvious advantage -

Chapter 9: AI-Assisted Design and Engineering

AI can generate and evaluate hundreds of design iterations in the time it would take a human to create just a few. But beyond efficiency, generative design enhances creativity by proposing unconventional solutions that push beyond traditional design boundaries. It also promotes sustainability through intelligent material optimization, reducing waste while maintaining structural integrity. The result is improved product performance, as AI-optimized designs often achieve better functionality than conventionally designed alternatives, leading to products that are lighter, stronger, and more efficient.

Real-world applications demonstrate generative design's transformative potential. In aerospace, companies like Boeing and Airbus use generative design to create components that are significantly lighter yet stronger than traditional designs, directly improving fuel efficiency and reducing environmental impact. The automotive industry has embraced this technology to accelerate development cycles - what once took months of iterative design can now be accomplished in weeks. Manufacturers report not only faster time-to-market but also more durable vehicle components that perform better under stress. These case studies illustrate how generative design delivers tangible business value across industries.

AI integration in CAD software represents a paradigm shift from passive design tools to intelligent design partners. Feature recognition capabilities allow AI to automatically identify and classify design elements, dramatically improving modeling efficiency and reducing human error. Error detection systems continuously monitor designs, flagging potential issues before they become costly problems in manufacturing or deployment. Perhaps most valuable are AI-powered design improvement suggestions that analyze current designs against best practices and optimization principles, offering recommendations that engineers might not have considered. This transforms CAD from a drafting tool into an intelligent design advisor.

AI's impact on simulation extends beyond simple automation to fundamental improvements in accuracy and speed. Predictive modeling capabilities allow AI to forecast simulation outcomes before running full calculations, enabling engineers to focus computational resources on the most promising scenarios. Parameter optimization algorithms fine-tune simulation settings automatically, ensuring maximum accuracy while minimizing computation time. Real-time analysis capabilities provide continuous monitoring and instant feedback, enabling rapid decision-making during the design process. This combination of speed and accuracy allows engineers to explore more design alternatives and make better-informed decisions faster than ever before.

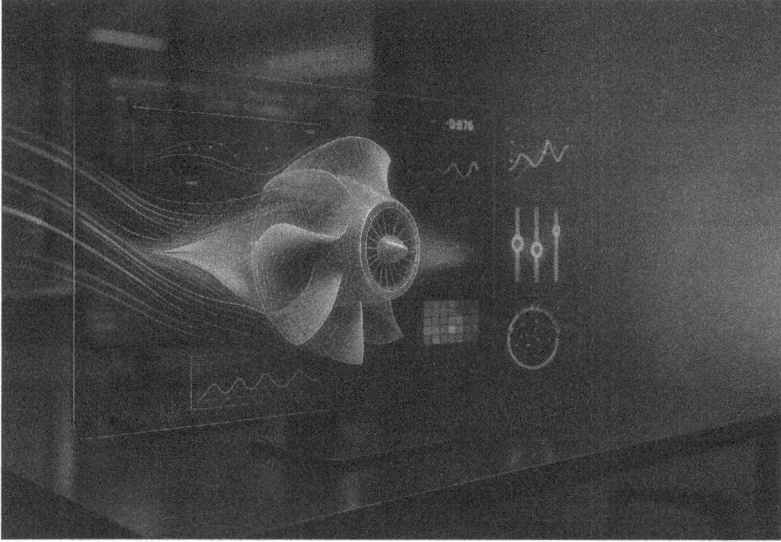

AI co-pilots are transforming daily engineering workflows in practical, measurable ways. By automating repetitive tasks like dimensioning, annotation, and routine calculations, they free engineers to focus on creative problem-solving and innovation. In complex assembly management, AI co-pilots track component relationships, detect interferences, and suggest optimization strategies that human engineers might miss. The quality improvements are substantial - AI systems can identify potential design flaws, manufacturing challenges, and performance bottlenecks early in the design process. This results in higher-quality final products and significantly reduced development costs through early problem identification and resolution.

Natural Language Processing is revolutionizing how engineers create and interact with technical documentation. Automated report generation transforms raw simulation data, test results, and design specifications into comprehensive, well-structured reports that follow industry standards. Key information extraction capabilities allow NLP systems to automatically identify and highlight critical data points, specifications, and requirements from lengthy technical documents. Document summarization creates concise executive summaries of complex technical reports, enabling faster decision-making and better communication across engineering teams. These capabilities dramatically reduce the time engineers spend on documentation while improving consistency and accuracy.

Knowledge management benefits from NLP through dramatic improvements in information accessibility and searchability. Structuring unstructured data transforms scattered technical information into organized, searchable databases that can be easily navigated. Semantic search capabilities go beyond keyword matching to understand the meaning and context of queries, returning more relevant results. Enhanced information retrieval systems help engineers quickly locate specific technical information, reducing research time and preventing the loss of valuable institutional knowledge. This is particularly crucial in large organizations where technical expertise and historical project knowledge can easily become fragmented or lost.

Processing technical language presents unique challenges that require specialized solutions. Engineering terminology is highly specialized, context-dependent, and constantly evolving, making it difficult for general-purpose NLP systems to process accurately. Domain-specific NLP models address these challenges by training algorithms specifically on engineering vocabulary, technical standards, and industry-specific language patterns. Advanced data annotation techniques ensure that these models understand the nuanced relationships between technical concepts. The result is dramatically improved accuracy in processing engineering documents, enabling more reliable automation of documentation tasks and better knowledge extraction from technical texts.

Mining valuable insights from legacy data requires sophisticated analytical techniques that can handle the unique challenges of historical engineering information. Data cleaning is the essential first step, removing inconsistencies, correcting errors, and standardizing formats across potentially decades of accumulated data. Pattern recognition algorithms identify meaningful trends, recurring failure modes, and successful design strategies that might be buried in vast datasets. Machine learning analysis detects anomalies that could indicate emerging problems or opportunities for improvement. These techniques transform legacy data from static historical records into actionable intelligence for current and future projects.

Integrating historical data into modern workflows creates powerful synergies between past experience and current capabilities. Enhanced simulations benefit from historical performance data, providing more accurate baseline conditions and validation benchmarks. Predictive maintenance programs leverage decades of equipment performance data to identify early warning signs and optimize maintenance schedules. New designs incorporate lessons learned from previous projects, helping engineers avoid repeating past mistakes while building on proven successful strategies. This integration ensures that valuable institutional knowledge is preserved and actively contributes to ongoing innovation rather than being lost to time or personnel changes.

While automated knowledge extraction offers significant benefits, understanding its limitations is crucial for successful implementation. Time-saving benefits are substantial - AI can process years of accumulated data in hours rather than months. Insight discovery capabilities often reveal hidden patterns that human analysts might miss. However, data quality challenges remain significant - poor quality input data will inevitably lead to unreliable results. Context interpretation limits mean that human validation remains essential for ensuring accuracy and relevance. The most successful implementations combine AI efficiency with human expertise, creating hybrid systems that leverage the strengths of both automated processing and human judgment.

AI-assisted design and engineering represents a fundamental transformation in how we approach innovation and problem-solving. Enhanced creativity emerges from the collaboration between human insight and AI computational power, enabling solutions that neither could achieve alone. Improved efficiency through intelligent automation allows engineers to focus on high-value creative work while AI handles routine tasks. Advanced knowledge management systems ensure that institutional wisdom is preserved, organized, and made accessible to current and future engineering teams. The future of engineering lies not in replacing human expertise with AI, but in creating powerful partnerships that amplify human capabilities and unlock new possibilities for innovation.

Chapter 10: Robotics and Human-Machine Collaboration

Collaborative robots, or cobots, represent a fundamental shift from traditional industrial robotics. Unlike conventional robots that operate in isolation, cobots are specifically engineered to work safely alongside humans in shared workspaces. Their enhanced safety features include force-limiting technology, collision detection sensors, and sophisticated programming that ensures immediate response to human presence. These robots excel in human-robot collaboration by taking on repetitive, dangerous, or precision-requiring tasks while humans focus on complex decision-making and creative problem-solving. Cobots bridge the gap between full automation and human capability, creating synergistic partnerships that enhance overall productivity.

Reinforcement learning transforms how cobots acquire and refine their skills through continuous interaction with their environment. This trial-and-error learning approach allows cobots to discover optimal behaviors without explicit programming for every scenario. Through iterative experimentation and feedback analysis, they develop improved adaptability, becoming more flexible in responding to unexpected situations or changing task requirements. The continuous feedback-driven learning process significantly enhances task performance optimization, enabling cobots to achieve greater efficiency and precision over time. This adaptive learning capability ensures that cobots can evolve their performance in dynamic manufacturing and service environments.

Chapter 10: Robotics and Human-Machine Collaboration

Integrating cobots into workplaces offers substantial benefits while presenting unique challenges that organizations must address strategically. The primary advantages include increased productivity through task automation and enhanced safety by removing humans from dangerous operations. However, successful integration faces several critical challenges: ensuring smooth human-robot interaction requires sophisticated interface design and worker training. Managing integration costs involves balancing initial investment against long-term productivity gains. Additionally, workforce adaptation presents cultural and technical challenges as employees learn to collaborate with robotic partners. Organizations must develop comprehensive change management strategies to maximize cobot benefits while addressing legitimate worker concerns about job displacement and role transformation.

Safety AI serves as the cornerstone of effective human-robot collaboration, establishing trust and preventing accidents in shared workspaces. These systems enable robots to continuously detect human presence and predict potential collision scenarios, allowing for immediate preventive action. The accident prevention capabilities rely on sophisticated sensors and algorithms that monitor the robot's operational environment in real-time. Building trust between humans and robots requires consistent, predictable safety responses that workers can rely upon. Dynamic environment adaptation ensures robots can safely navigate changing conditions, varying human movements, and unexpected obstacles. Effective safety AI creates confident human-robot partnerships essential for productive collaboration.

Vision-based obstacle detection employs multiple complementary technologies to create comprehensive environmental awareness for robotic systems. Camera-based detection captures rich visual data, enabling robots to identify and classify obstacles, humans, and objects in complex three-dimensional environments. LIDAR technology provides precise distance measurements and detailed 3D mapping capabilities, creating accurate spatial representations for navigation planning. The integration of multiple sensor types—including cameras, LIDAR, ultrasonic sensors, and infrared detection—creates redundant safety systems that enhance reliability. This sensor fusion approach ensures robust real-time obstacle detection, enabling safer robot movements around humans and objects while maintaining operational efficiency in dynamic work environments.

Real-world safety implementations demonstrate the practical value of AI-driven safety systems across diverse industries. In manufacturing environments, safety AI has measurably reduced workplace incidents while simultaneously enhancing operational efficiency through optimized robot-human coordination. Service robots operating in public spaces, hospitals, and offices utilize integrated safety AI to navigate complex human environments while maintaining appropriate social distancing and interaction protocols. Human-machine workspace designs incorporating comprehensive safety measures have proven to boost productivity significantly while minimizing accident risks. These case studies validate that properly implemented safety AI creates win-win scenarios where both safety and efficiency improve simultaneously through intelligent collaboration design.

Gesture recognition systems transform human-machine interaction by interpreting natural hand and body movements as control commands. These systems analyze gesture interpretation through sophisticated computer vision algorithms that track movement patterns and translate them into meaningful machine instructions. Advanced cameras and sensors capture detailed motion data, enabling machines to understand and respond to intuitive human gestures naturally. The resulting touchless interaction eliminates the need for physical interfaces, reducing contamination risks and improving ergonomics. This technology enhances user experience by making machine control more intuitive and accessible, particularly valuable in environments where hands-free operation is essential, such as medical settings or manufacturing environments requiring sterile conditions.

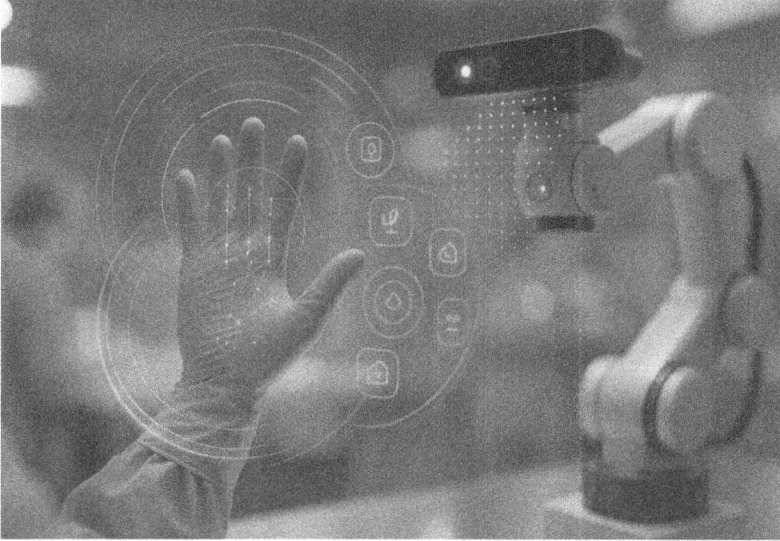

Voice command technologies create intuitive control interfaces by combining speech recognition with natural language processing capabilities. Speech recognition technology converts spoken words into digital commands that machines can interpret and execute accurately and efficiently. Natural language processing enables systems to understand context, intent, and conversational nuances, allowing for more sophisticated human-machine communication beyond simple command structures. The hands-free operation benefits extend beyond convenience to significant productivity and safety improvements. Workers can control machines while maintaining focus on their primary tasks, keeping hands free for manual operations, and avoiding potential contamination from shared control surfaces. This technology proves particularly valuable in fast-paced or sterile work environments.

Multimodal interfaces combining gesture and voice controls create flexible, adaptive systems that accommodate diverse user preferences and abilities. This multimodal interaction approach recognizes that different users may prefer different control methods based on their physical capabilities, cultural backgrounds, or situational contexts. Improved ergonomics result from allowing users to choose the most comfortable and natural interaction method, reducing physical strain and fatigue associated with repetitive manual controls. The accessibility benefits are significant, as multimodal interfaces can accommodate users with various physical limitations or preferences. By providing multiple interaction pathways, these systems ensure that advanced technology remains accessible and usable across diverse user populations.

Adaptive task allocation algorithms represent sophisticated decision-making systems that optimize human-machine collaboration through intelligent work distribution. These systems begin with comprehensive assessment of capabilities, evaluating both human strengths—creativity, problem-solving, adaptability—and machine advantages—precision, endurance, computational power. Effective workload management prevents both human overload and machine underutilization by dynamically balancing task assignments based on current capacity and performance metrics. Context awareness enables these algorithms to adapt task allocation according to changing environmental conditions, urgency levels, and available resources. The ultimate goal is effective collaboration that maximizes combined human-machine team performance while ensuring sustainable workload distribution and job satisfaction.

Real-time optimization systems continuously monitor and adjust human-machine team performance to maintain peak efficiency and responsiveness. Continuous monitoring involves tracking multiple performance indicators: human fatigue levels, machine operational status, task completion rates, and quality metrics. This comprehensive data collection enables intelligent decision-making about task redistribution and resource allocation. Dynamic task assignment responds to changing conditions by automatically reassigning tasks based on current team capabilities and priorities. For example, if a human team member shows fatigue indicators, the system can shift more routine tasks to machines while preserving human involvement in creative or complex problem-solving activities. This adaptive approach maximizes team productivity while maintaining worker satisfaction.

The future of adaptive collaboration promises enhanced efficiency, worker satisfaction, and technological integration that will transform modern workplaces. Improved efficiency and flexibility through adaptive task allocation enable teams to respond more effectively to changing demands and unexpected challenges. Worker satisfaction increases when tasks align with individual strengths and preferences, creating more engaging and fulfilling work experiences. AI-driven predictive models represent the next evolution, using machine learning to anticipate optimal task allocation patterns before bottlenecks occur. Wearable technology integration will provide real-time biometric feedback, stress indicators, and performance data, enabling even more sophisticated and responsive adaptive collaboration systems that optimize both productivity and human wellbeing.

This chapter has explored three transformative areas reshaping human-machine collaboration. Advances in cobots demonstrate how collaborative robots enhance workplace safety and efficiency through seamless human-robot partnerships. AI safety measures and intuitive gesture and voice interfaces make human-machine interaction more natural, secure, and accessible to diverse users. Adaptive algorithms enable systems to continuously learn and optimize collaboration patterns, leading to enhanced productivity and more effective teamwork. Together, these innovations point toward a future where humans and machines work as integrated teams, each contributing their unique strengths to achieve outcomes neither could accomplish alone. The key to success lies in thoughtful implementation that prioritizes both efficiency and human-centered design.

Chapter 11: AI for Skills Development and Training

AI is fundamentally transforming skills enhancement through three primary mechanisms. Personalized learning paths represent the most significant advancement, where AI algorithms analyze individual learner behavior, preferences, and performance to create customized educational experiences. This approach ensures that each learner receives content tailored to their specific needs and learning pace. Automated assessments eliminate human bias and provide consistent, objective evaluation of learner progress. These systems can process vast amounts of data to identify patterns and provide detailed feedback. Data-driven training optimization uses analytics to continuously refine and improve training programs, ensuring they remain effective and relevant to changing industry needs and learner requirements.

AI-driven training solutions offer three critical benefits that traditional methods struggle to achieve. Enhanced engagement occurs through interactive and personalized learning experiences that maintain learner interest and motivation. These systems adapt content presentation styles, difficulty levels, and pacing to match individual preferences, resulting in significantly higher completion rates and satisfaction scores. Scalable customization allows organizations to deliver personalized training to hundreds or thousands of learners simultaneously without proportional increases in human resources. Advanced analytics provide unprecedented insights into learner performance, identifying knowledge gaps, predicting potential difficulties, and enabling proactive interventions. These analytics also help organizations optimize their training investments and demonstrate measurable returns on educational initiatives.

Despite their benefits, AI adoption in training faces four significant challenges that organizations must address. Data privacy concerns are paramount as training systems collect and analyze sensitive learner information. Organizations must implement robust security measures and comply with regulations like GDPR to maintain learner trust. System integration challenges arise when implementing AI solutions within existing training infrastructure, requiring technical expertise and careful planning to ensure seamless operation. Algorithmic fairness is crucial to prevent bias and ensure equitable access to learning opportunities across diverse learner populations. Change management becomes essential as organizations must prepare staff and learners for new AI-enhanced training methods while addressing resistance to technological change and ensuring smooth transitions.

Chapter 11: AI for Skills Development and Training

Digital twins represent a revolutionary approach to training simulation by creating precise virtual replicas of real-world systems or environments. Real-time data integration forms the foundation, with digital twins continuously receiving live data feeds from sensors, databases, and other sources to maintain accuracy and relevance. This ensures that the virtual environment reflects current conditions and responds realistically to learner interactions. Three-dimensional modeling and simulation provide immersive visualization capabilities that enable learners to explore and interact with complex systems safely. AI-driven interaction enhances these digital twins by incorporating predictive analytics, intelligent responses, and adaptive scenarios that respond dynamically to learner actions and decisions, creating truly interactive training environments.

Digital twins excel in three critical training applications where real-world practice would be costly, dangerous, or impractical.

Manufacturing training benefits enormously from digital twin simulations that replicate complex production processes, allowing workers to practice equipment operation, troubleshoot problems, and learn safety procedures without risking expensive machinery or production downtime. Healthcare education leverages digital twins to provide medical students and professionals with realistic patient scenarios and surgical simulations, enabling them to develop clinical skills in a risk-free environment. Aviation training has pioneered digital twin technology through sophisticated flight simulators that recreate various aircraft and weather conditions, allowing pilots to experience and respond to emergency situations safely while building confidence and competence.

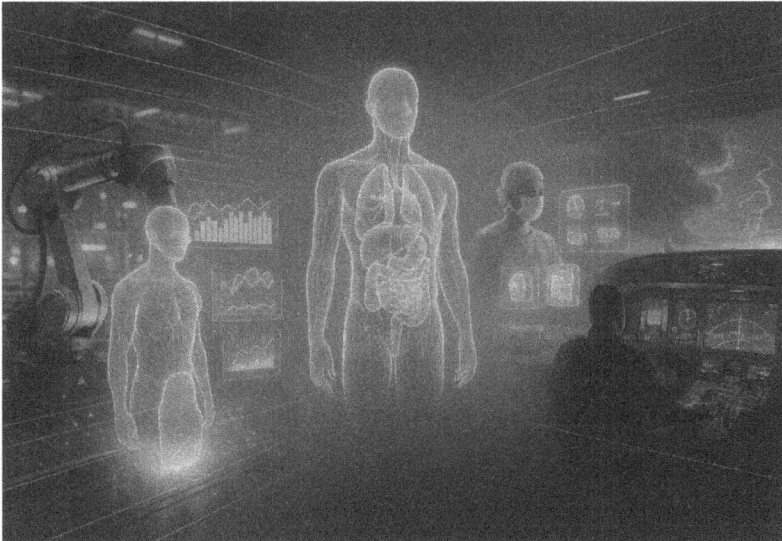

Digital twins offer four distinct advantages over traditional simulation methods. Enhanced accuracy stems from their ability to replicate real-world systems more precisely by incorporating actual system data, physics-based modeling, and real-time updates. This accuracy translates to better skill transfer and more effective training outcomes. Real-time updates ensure that the training environment reflects current conditions and system status, providing learners with the most relevant and up-to-date experience possible. Personalized scenarios allow trainers to create specific situations tailored to individual learner needs, skill levels, and learning objectives. Improved skill transfer occurs because digital twins closely mimic real-world dynamics and environments, helping learners apply their training more effectively when transitioning to actual work situations.

AI agents revolutionize learning experiences through four key personalization mechanisms. Behavior analysis involves continuous monitoring of learner actions, preferences, response times, and engagement patterns to build comprehensive learner profiles. This data enables AI agents to understand how individuals learn best and what content resonates with them. Adaptive difficulty dynamically adjusts challenge levels based on real-time assessment of learner progress and capabilities, ensuring optimal learning zones that prevent frustration while maintaining engagement. Interactive dialogues create natural language conversations between learners and AI agents, providing explanations, answering questions, and offering guidance in an intuitive, conversational manner. Enhanced motivation results from this personalization, as learners experience content that feels relevant and appropriately challenging, leading to improved knowledge retention and skill development.

Adaptive learning systems implement three essential mechanisms for effective progression tracking. Continuous progress monitoring involves real-time assessment of learner performance across multiple dimensions, including comprehension, skill application, and engagement levels. This constant feedback loop enables immediate identification of learning difficulties and successes. Adaptive goal adjustment dynamically modifies learning objectives based on individual progress, ensuring that goals remain challenging yet achievable while addressing specific areas needing improvement. The system prioritizes skill gaps and adjusts the learning path accordingly. Efficient skill acquisition results from this personalized approach, as learners receive targeted practice in areas where they need the most support, while spending less time on concepts they've already mastered, optimizing both learning time and outcomes.

Chapter 11: AI for Skills Development and Training

Real-world implementations of personalized training agents demonstrate significant success across multiple domains. AI tutors in language learning have transformed how people acquire new languages by providing personalized conversation practice, grammar correction, and cultural context. These systems adapt to individual speaking patterns, learning pace, and cultural background to optimize language acquisition. Virtual coaches in corporate training deliver personalized guidance for professional skill development, leadership training, and compliance education, adjusting content and delivery methods to match individual career paths and learning preferences. The benefits of personalization are measurable: organizations report increased learner engagement rates of 40-60%, faster skill mastery by 25-30%, and improved knowledge retention scores compared to traditional training methods. These improvements translate to reduced training costs and enhanced job performance.

Real-time monitoring in AI training systems relies on three sophisticated data collection mechanisms. Sensors capture continuous streams of learner actions, physiological responses, and environmental conditions during training sessions. These might include motion sensors, eye-tracking devices, heart rate monitors, and pressure sensors depending on the training context. Camera integration provides visual data for analyzing learner behavior, facial expressions, body language, and engagement levels. Computer vision algorithms process this visual information to assess attention, confusion, confidence, and emotional state. Data analytics for assessment processes all collected information using machine learning algorithms to instantly evaluate skills, identify errors, and assess performance against learning objectives. This comprehensive monitoring enables immediate intervention when learners struggle and recognition when they excel.

Immediate error identification and correction represent critical components of effective AI training systems. Automatic error detection utilizes advanced algorithms to instantly recognize mistakes, incorrect procedures, or suboptimal performance during training exercises. These systems compare learner actions against established best practices and immediately flag deviations. The speed of detection is crucial because it prevents learners from practicing and reinforcing incorrect techniques that could become difficult to correct later. Corrective instructions provide immediate guidance through various modalities including visual cues, audio feedback, haptic responses, or interactive demonstrations. This real-time correction helps learners understand not just what they did wrong, but why it was incorrect and how to perform the task correctly, facilitating faster learning and skill development.

Real-time feedback significantly impacts learner outcomes through three key psychological and educational mechanisms. Boosting learner confidence occurs when systems provide immediate positive reinforcement for correct actions and gentle correction for mistakes, helping learners feel supported throughout their learning journey. This immediate feedback reduces anxiety and builds self-efficacy. Reducing learner frustration happens because confusion and errors are addressed immediately rather than allowing learners to struggle with misunderstandings that could compound over time. When learners know they're making progress and understand their mistakes quickly, they maintain motivation and engagement. Enhanced skill retention results from the immediate connection between action and feedback, which strengthens neural pathways and improves long-term memory formation. This timely feedback loop creates more effective learning experiences and better practical application of acquired skills.

Knowledge graphs provide a sophisticated framework for organizing and connecting information in AI systems. Entities and relationships form the basic structure, where nodes represent individual concepts, people, objects, or ideas, while edges represent the connections and relationships between them. This structure mirrors how human knowledge is interconnected rather than stored in isolated silos. Semantic understanding emerges from this graph structure, enabling AI systems to comprehend not just individual pieces of information but the contextual relationships and meanings between different concepts. This understanding allows for more nuanced and intelligent responses to queries. Supporting complex queries becomes possible because knowledge graphs can trace multiple relationship paths to answer sophisticated questions that require understanding connections across different domains of knowledge, providing comprehensive and contextually relevant information.

Integration of knowledge graphs with AI expert systems creates powerful capabilities for training applications. Advanced reasoning emerges when expert systems can leverage the rich relationship data in knowledge graphs to make more sophisticated inferences and decisions. This combination allows systems to consider multiple factors, historical context, and complex interdependencies when providing guidance or making recommendations. Personalized recommendations become more accurate and relevant because the system understands not only learner preferences but also the relationships between different concepts, skills, and learning objectives. Enhanced problem-solving capabilities result from the system's ability to draw upon vast networks of connected knowledge to suggest solutions, identify patterns, and provide comprehensive support for complex training scenarios. This integration creates more intelligent and capable training systems that can adapt to diverse learning needs.

Knowledge graphs significantly enhance decision support and knowledge transfer in training environments. Rapid expertise access becomes possible as systems can quickly traverse knowledge networks to locate and present relevant expert information, best practices, and specialized knowledge exactly when learners need it. This eliminates delays in finding critical information during training exercises. Improved learning pathways result from the system's ability to understand prerequisite relationships, skill dependencies, and optimal learning sequences. The knowledge graph helps identify the most efficient path through complex subject matter based on individual learner needs and existing knowledge. Preserving institutional knowledge ensures that valuable organizational expertise doesn't disappear when experienced employees retire or leave. Knowledge graphs capture and maintain these insights in a structured, accessible format that can continue to benefit future learners and training programs.

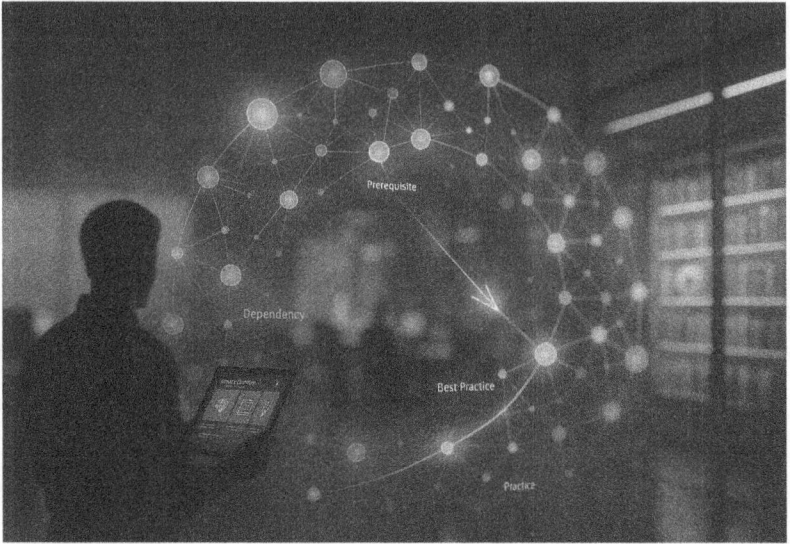

In conclusion, this chapter has explored four transformative AI applications in skills development and training. Digital twins in training create immersive, risk-free environments that simulate real-world conditions with unprecedented accuracy and real-time data integration. Personalized learning agents revolutionize individual learning experiences by adapting content, difficulty, and presentation to match each learner's unique needs and preferences. Real-time feedback systems provide immediate insights and corrections that enhance performance, reduce frustration, and improve skill retention through timely intervention. Knowledge graphs organize complex information networks to support adaptive, contextualized learning experiences that leverage interconnected knowledge for more effective training outcomes. These technologies collectively represent the future of skills development, offering scalable, effective, and engaging training solutions.

Chapter 11: AI Governance and Compliance

AI governance in industrial environments serves as the foundation for safe, ethical, and effective AI implementation. The primary purpose is to ensure AI systems operate reliably within the complex constraints of industrial operations. Key components include establishing clear policies that define acceptable AI use, implementing robust processes for AI development and deployment, creating control mechanisms for oversight, defining accountability structures, and maintaining continuous oversight. Organizational alignment is crucial - AI governance must seamlessly integrate with existing business objectives while meeting all relevant compliance requirements. This alignment ensures that AI investments support strategic goals rather than creating operational conflicts or regulatory violations.

Chapter 11: AI Governance and Compliance

Industrial AI implementation faces a complex web of compliance requirements across multiple dimensions. International standards provide the global framework for interoperability and ethical AI use, ensuring systems can operate across borders while maintaining consistent quality and safety standards. Industry-specific regulations address unique sector requirements - manufacturing, healthcare, finance, and other industries each have tailored safety and quality expectations that AI systems must meet. Data protection laws, including GDPR, CCPA, and similar frameworks, are critical for protecting user information and building stakeholder trust. Finally, organizations must develop comprehensive internal policies that guide ethical AI deployment, establish clear usage guidelines, and ensure responsible implementation across all levels of the organization.

Maintaining governance standards in rapidly evolving AI environments presents several significant challenges. Rapid technology evolution creates a moving target - as AI capabilities advance quickly, governance frameworks struggle to keep pace with new technologies and their implications. Complex global supply chains introduce multiple points of potential failure or non-compliance, making it difficult to maintain consistent standards across all vendors and partners. Legacy systems integration poses technical challenges when implementing new governance standards alongside existing industrial infrastructure that may not have been designed for AI oversight. Perhaps most critically, ensuring adequate human oversight in increasingly autonomous AI applications requires careful balance between efficiency gains and maintaining ethical governance standards throughout all automated processes.

Transparent AI models are essential for building stakeholder confidence and ensuring effective industrial operations. Enhanced stakeholder trust emerges when decision processes are clear and understandable, enabling operators, managers, and regulators to have confidence in AI-driven outcomes. Promoting accountability becomes possible when AI outputs are transparent, allowing operators and regulators to hold systems accountable for their decisions and understand the rationale behind critical choices. Supporting effective decision-making is perhaps the most practical benefit - transparent AI models provide actionable insights that improve human decision-making in complex industrial environments, enabling better collaboration between human expertise and artificial intelligence capabilities for optimal operational outcomes.

Several techniques can enhance explainability in AI systems, each serving different needs and technical constraints. Model-agnostic methods like LIME (Local Interpretable Model-agnostic Explanations) and SHAP (SHapley Additive exPlanations) provide explanations regardless of the underlying AI model architecture, increasing flexibility and enabling consistent explanation approaches across diverse AI implementations. Interpretable models such as decision trees, linear regression, and rule-based systems offer inherently transparent decision-making processes that humans can easily understand and validate. Visualization tools serve as the bridge between complex AI computations and human comprehension, using intuitive graphics, charts, and interactive displays to clarify AI decision processes and make complex patterns accessible to non-technical stakeholders.

Transparency directly impacts trust and adoption rates across multiple dimensions of industrial AI implementation. Fostering user confidence occurs when transparency makes AI decisions clear and understandable, reducing the "black box" effect that often creates hesitation among users and stakeholders. Promoting regulatory compliance becomes more achievable when transparent AI systems can demonstrate adherence to legal standards and regulatory requirements through clear documentation of decision processes. Reducing resistance to adoption addresses one of the biggest barriers to AI implementation - users are significantly more likely to embrace AI systems when transparency minimizes fear and uncertainty about automated decision-making. Improving human-AI collaboration creates synergistic relationships where transparency enables better cooperation between human expertise and AI capabilities for more effective outcomes.

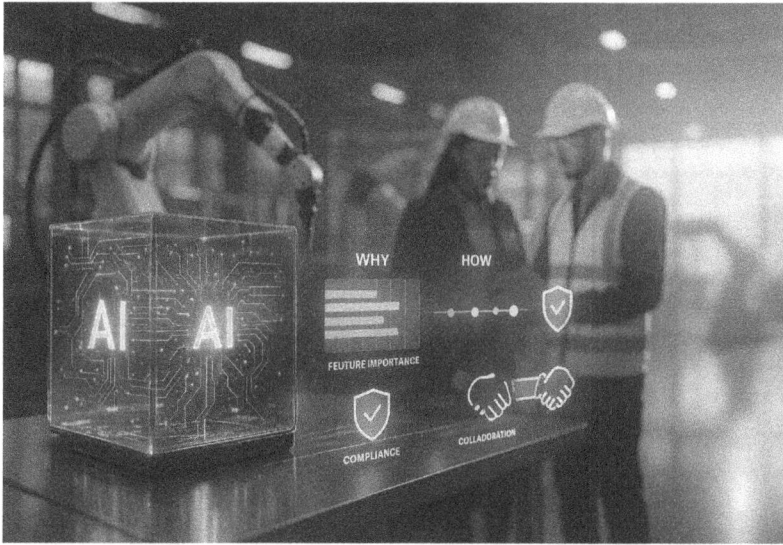

Industrial AI applications face unique risks that require specialized identification and management approaches. System failures in industrial contexts can cause unexpected downtime, compromising operational efficiency and potentially leading to significant financial losses or safety incidents. Cybersecurity threats represent a critical vulnerability as AI systems become attractive targets for malicious actors seeking to breach data security or attack critical infrastructure. Data biases present ongoing challenges that can lead to incorrect decisions, potentially impacting both safety protocols and productivity metrics in ways that compound over time. Operational disruptions caused by AI malfunctions or errors can halt production lines, create cascading delays throughout manufacturing processes, and affect entire supply chains, making robust risk management essential for industrial AI success.

Several established frameworks provide structured approaches to AI risk management in industrial contexts. The ISO 31000 framework offers comprehensive principles and guidelines for risk management that can be effectively applied across various industries and AI implementation contexts, providing a solid foundation for systematic risk assessment. The NIST AI Risk Management Framework specifically addresses AI-related risks, guiding organizations through the process of assessing, mitigating, and managing risks unique to artificial intelligence systems. Industry-specific guidelines recognize that different sectors face unique challenges and provide tailored best practices for managing AI risks within particular industrial contexts, ensuring that risk management approaches align with sector-specific safety requirements, regulatory frameworks, and operational constraints.

Effective risk mitigation requires continuous improvement strategies that adapt to evolving industrial needs and technological developments. Ongoing monitoring involves tracking AI system performance in real-time to identify potential risks early and maintain control effectiveness before issues escalate into operational problems. Regular feedback loops enable systematic adaptation and refinement of risk controls based on real-world outcomes and operational experience, ensuring that mitigation strategies evolve with changing conditions. Periodic audits and updates ensure that risk mitigation strategies remain aligned with evolving industrial needs, regulatory changes, and technological advances. This systematic approach creates a robust framework for maintaining AI system reliability and safety over time while adapting to new challenges and opportunities.

Chapter 11: AI Governance and Compliance

The current regulatory landscape for AI in manufacturing encompasses multiple overlapping frameworks that organizations must navigate carefully. Data protection laws regulate the handling of personal and sensitive information in AI manufacturing processes globally, with frameworks like GDPR setting strict requirements for data collection, processing, and storage. Product safety standards ensure that AI-driven manufacturing outputs meet rigorous safety and quality requirements, maintaining consumer protection and industry standards. AI-specific guidelines focus on ethical AI use, transparency requirements, and accountability measures in manufacturing technologies, addressing the unique challenges posed by automated decision-making. Industry-specific mandates recognize that different sectors enforce unique regulatory requirements tailored to their particular manufacturing AI applications, creating a complex but necessary regulatory environment.

Regulatory compliance presents several significant challenges that organizations must address proactively. Ambiguous and evolving regulations often lack clarity and change continuously, creating uncertainty for compliance efforts and making it difficult to establish stable, long-term compliance strategies. Cross-jurisdictional differences create complexity for organizations operating internationally, as diverse regulations across regions complicate compliance efforts and require sophisticated legal and operational coordination. Applying traditional regulatory models to AI technologies presents fundamental challenges, as existing regulatory frameworks often struggle to address the unique aspects of AI technologies, including machine learning adaptability, autonomous decision-making, and the dynamic nature of AI system behavior over time.

Successful alignment of AI systems with regulatory requirements requires proactive strategies that address compliance from the design phase through implementation. Regulatory impact assessments conducted early in AI system development help identify compliance risks before they become expensive problems, enabling organizations to make informed decisions about system design and implementation approaches. Stakeholder engagement ensures diverse perspectives are considered and helps achieve alignment with regulatory expectations by involving relevant parties in the development process. Transparent design and thorough documentation increase accountability and support regulatory compliance by creating clear audit trails and demonstrating adherence to required standards. Compliance automation tools streamline ongoing compliance monitoring and reporting for AI systems, reducing manual oversight burden while maintaining consistent compliance standards.

Intellectual property issues in AI development present complex challenges that require careful legal and strategic consideration. Patent eligibility remains a key challenge as determining whether AI inventions qualify for patent protection involves complex legal questions about inventorship, novelty, and the role of human versus machine contribution to innovations. Copyright issues arise regarding works created by AI systems without direct human authorship, creating uncertainty about ownership and protection of AI-generated content. Trade secrets protection becomes critical for maintaining competitive advantage, as AI algorithms and models often represent valuable proprietary information that requires careful protection strategies. Licensing agreements play an essential role in regulating the use and distribution of AI algorithms and models between parties, requiring careful negotiation of rights and responsibilities.

Data ownership and privacy concerns represent fundamental challenges in industrial AI implementation that require comprehensive management strategies. Sensitive data handling becomes critical as AI systems process proprietary and confidential information, requiring strict protocols to protect user information, maintain privacy, and preserve competitive advantages. Consent and privacy laws, including frameworks like GDPR, regulate how personal data can be collected, processed, and used in AI technologies, requiring organizations to implement robust systems for managing consent and ensuring compliance with individual privacy rights. Data stewardship ensures responsible management and ethical use of data in AI applications, establishing clear governance frameworks for data collection, processing, storage, and sharing throughout the AI lifecycle.

Effective protection of intellectual property and data requires implementing comprehensive best practices across multiple security and legal dimensions. Robust contracts form the foundation by clearly defining intellectual property rights, data usage policies, and responsibilities of all parties involved in AI development and deployment. Access controls and encryption provide technical safeguards by implementing strict access management and data encryption protocols that ensure confidentiality and prevent unauthorized access to sensitive information and proprietary algorithms. Audit trails and compliance documentation support transparency and help organizations demonstrate adherence to privacy regulations and industry standards, creating accountability frameworks that protect both intellectual property and stakeholder interests while enabling effective oversight and risk management.

Successful AI implementation in industrial contexts requires a holistic governance framework that integrates multiple critical elements. A holistic governance framework supports responsible AI deployment by combining policy, technical, and operational considerations into a coherent management approach. Transparency and risk management prove essential for ensuring safe and effective AI operations, providing the foundation for stakeholder trust and regulatory compliance. Regulatory and IP compliance safeguards AI implementation by protecting organizational interests while meeting legal requirements

136

and ethical standards. Organizations that successfully integrate these elements create sustainable AI implementations that deliver business value while maintaining safety, compliance, and ethical operation standards.

Chapter 13: Scaling AI in Manufacturing

The AI Centre of Excellence serves as your organization's strategic hub for advancing artificial intelligence initiatives across manufacturing operations. This dedicated team functions as the central nervous system for AI adoption, ensuring responsible implementation while establishing robust governance frameworks that align with your business objectives. The CoE goes beyond just technology deployment - it focuses on developing critical AI skills and capabilities throughout your organization. This holistic approach ensures that your manufacturing teams have both the technical expertise and strategic guidance needed to leverage AI effectively, creating a sustainable foundation for long-term digital transformation success.

A successful AI Centre of Excellence operates across five key functional areas that collectively drive organizational transformation. Setting AI standards ensures consistent, effective implementation across all manufacturing processes, preventing fragmented approaches that can undermine success. Project management capabilities ensure AI initiatives align with organizational goals and deliver on time and within budget. The CoE actively fosters collaboration between AI developers and manufacturing experts, breaking down silos that often impede innovation. Talent development through comprehensive training programs builds internal expertise and reduces dependency on external resources. Finally, ethical AI usage ensures all applications comply with regulatory requirements and ethical standards, protecting your organization's reputation and maintaining stakeholder trust.

Chapter 13: Scaling AI in Manufacturing

Establishing and scaling an effective AI Centre of Excellence requires four critical success factors. Executive sponsorship provides the strategic direction and resource allocation necessary for meaningful impact - without C-level commitment, CoEs often struggle to achieve their full potential. Cross-functional involvement ensures diverse expertise and promotes collaboration across departments, preventing the CoE from becoming an isolated technology function. Clear governance frameworks guide decision-making processes and establish accountability structures that enable effective oversight and risk management. Continuous learning and scalability allow the CoE to adapt to evolving technologies and business needs while expanding its influence across the organization. These elements work synergistically to create a robust foundation for AI excellence.

AI maturity in manufacturing rests on four foundational pillars that must be developed systematically. Robust data infrastructure serves as the backbone, enabling efficient collection, processing, and analysis of manufacturing data across all operational systems. A skilled workforce is essential for implementing and managing AI technologies effectively - this includes both technical specialists and manufacturing professionals who understand how to integrate AI into existing processes. Clear governance policies ensure responsible AI adoption while maintaining regulatory compliance and managing risks appropriately. Finally, technology adoption must be coupled with a culture that embraces innovation, as technical capabilities alone cannot drive successful transformation without organizational readiness and change management.

AI maturity typically progresses through four distinct stages, each building upon the previous level. Initial experimentation involves small-scale AI projects designed to test feasibility and demonstrate potential benefits to stakeholders. This stage focuses on learning and building confidence rather than immediate ROI. Pilot projects represent the next evolution, validating AI solutions in controlled environments before broader implementation. These projects provide crucial insights into integration challenges and help refine approaches. The integration stage sees AI solutions incorporated into existing systems to improve specific processes and workflows. Finally, enterprise-wide deployment represents full maturity, with AI implemented across the entire organization, supported by comprehensive governance structures and established performance metrics for continuous optimization.

Advancing AI maturity requires focused investment across four key areas. Investing in data quality is foundational - high-quality, well-structured data is essential for AI algorithms to function effectively and deliver reliable insights. Poor data quality will undermine even the most sophisticated AI implementations. Upskilling staff through comprehensive training programs enhances both AI adoption rates and operational efficiency, creating internal champions who can drive further innovation. Implementing robust governance mechanisms ensures responsible and effective AI deployment while managing risks and maintaining compliance. Leveraging pilot successes builds organizational momentum and secures stakeholder buy-in for broader implementation. These approaches should be pursued simultaneously rather than sequentially to maximize impact and accelerate maturity development.

Chapter 13: Scaling AI in Manufacturing

Build versus buy decisions for AI solutions require careful evaluation of four critical factors. Organizational expertise assessment determines whether your internal team has the skills and knowledge necessary to develop effective solutions - this includes both AI/ML capabilities and deep manufacturing domain knowledge. Time to market considerations influence whether internal development or commercial solutions better meet your urgency requirements. Customization needs vary significantly across manufacturing operations - highly specialized processes may require custom development, while standard applications might be adequately served by commercial solutions. Control and alignment requirements affect how closely the solution must integrate with your specific manufacturing goals and existing systems. These factors must be evaluated holistically rather than individually.

Evaluating build versus buy options requires systematic assessment across three key dimensions. Assessment of internal skills involves honest evaluation of your team's capabilities for solution implementation and ongoing management - consider both current skills and ability to acquire necessary expertise within required timeframes. Vendor solution features must be analyzed comprehensively, focusing on functionality, integration capabilities, and alignment with your specific operational needs rather than generic feature lists. Support and scalability considerations are crucial for long-term success - evaluate vendor support services, their track record with similar implementations, and the solution's ability to scale with your manufacturing operations. This evaluation should include both technical scalability and vendor capacity to support your growth.

The build versus buy decision involves fundamental trade-offs between investment, speed, and flexibility. Building solutions requires higher upfront investment and longer development timelines but offers superior customization and control over functionality. This approach is best suited for organizations with unique requirements, strong internal capabilities, and longer implementation timeframes. Buying solutions enables faster deployment and lower initial costs, making it attractive for organizations needing quick wins or lacking internal development capacity. However, commercial solutions may restrict customization options and flexibility for future modifications. The optimal choice depends on your specific circumstances, including timeline pressures, budget constraints, internal capabilities, and the degree of customization required for your manufacturing processes.

Calculating ROI for AI projects in manufacturing involves four primary value categories. Cost savings evaluation focuses on quantifiable reductions achieved through AI-driven process optimization, including labor cost reduction, energy efficiency improvements, and waste minimization. Productivity gains encompass the value of increased throughput, reduced cycle times, and error reduction in manufacturing workflows. Quality improvements deliver value through reduced defect rates, enhanced product consistency, and improved customer satisfaction scores. Revenue impacts include benefits from faster time-to-market, enhanced product offerings, and new revenue streams enabled by AI capabilities. These categories should be measured systematically using consistent methodologies to ensure accurate ROI calculations and meaningful comparisons across different AI initiatives.

Value realization from AI investments encompasses both tangible and intangible benefits that must be identified and tracked systematically. Monetary value represents the directly measurable financial benefits including cost reductions, productivity improvements, and revenue increases that can be quantified in traditional financial terms. Enhanced decision-making capabilities represent intangible value through improved choices based on better information and insights, leading to superior strategic outcomes. Risk reduction minimizes uncertainties and potential losses, adding significant value that may not be immediately apparent in financial statements. Innovation acceleration and workforce empowerment contribute to lasting competitive advantages and organizational capabilities that generate long-term benefits beyond immediate financial returns. Both categories should be tracked and communicated to stakeholders.

Chapter 13: Scaling AI in Manufacturing

Maximizing long-term benefits from AI investments requires systematic approaches to continuous improvement. Performance monitoring involves ongoing tracking of key metrics to assess progress and identify opportunities for optimization across all manufacturing processes. This should include both technical performance metrics and business impact measures. Feedback loops ensure continual refinement through regular communication between stakeholders and systematic assessment of results against objectives. Iterative enhancements implement incremental improvements based on learning and changing requirements, ensuring solutions evolve with your business needs. Alignment with evolving priorities ensures AI strategies adapt to changing manufacturing goals and market conditions. This comprehensive approach to continuous improvement ensures that AI investments continue delivering increasing value over time rather than diminishing returns.

This chapter has covered four essential elements for successfully scaling AI in manufacturing. Capability development through AI Centres of Excellence provides the organizational foundation necessary to support widespread AI adoption and ensure sustainable success. Maturity assessment helps organizations understand their current readiness level and identify specific areas requiring development before large-scale implementation. Strategic build versus buy decisions ensure optimal resource allocation and solution selection that aligns with organizational capabilities and requirements. Rigorous value measurement enables organizations to track both tangible and intangible benefits, ensuring AI investments deliver transformative results. Together, these elements create a comprehensive framework for scaling AI successfully across manufacturing operations, driving operational efficiency, and maintaining competitive advantage in an increasingly digital landscape.

Chapter 14: Future Trends and Disruptive Innovations

Foundation Models represent one of the most significant advances in industrial AI applications. These sophisticated AI systems, trained on vast datasets, offer unprecedented versatility in addressing complex manufacturing challenges. Unlike traditional narrow AI systems designed for specific tasks, foundation models can be adapted and fine-tuned for multiple manufacturing applications, from predictive maintenance to quality control. Their ability to understand and process complex industrial data makes them invaluable tools for modern manufacturers seeking to leverage AI for competitive advantage.

Foundation models serve as versatile AI frameworks that can be applied across diverse manufacturing scenarios. Their strength lies in their broad training on extensive datasets, which enables them to understand complex industrial contexts and relationships. These models excel at interpreting intricate industrial data patterns, from sensor readings to production metrics, enhancing both insight generation and analytical capabilities. Through fine-tuning processes, manufacturers can customize these models for specific tasks, improving operational efficiency and enabling more informed decision-making across their production systems.

Two primary use cases demonstrate the transformative power of foundation models in manufacturing: predictive maintenance and quality control enhancement. In predictive maintenance, these models analyze equipment data patterns to forecast potential failures before they occur, enabling optimized maintenance scheduling and significantly reducing unexpected downtime costs. For quality control, foundation models provide real-time anomaly detection capabilities, identifying product defects and process variations early in the production cycle. This early detection minimizes production issues, reduces waste, and ensures consistent product quality standards.

The integration of foundation models with Industrial Internet of Things (IoT) systems creates powerful synergies for manufacturing optimization. These models can process and analyze real-time sensor data from connected industrial devices, providing enhanced analytical capabilities that go beyond simple data collection. This integration enables instant processing of vast amounts of sensor data, supporting timely decision-making and operational improvements. Automated systems can respond quickly to AI-generated insights, enhancing manufacturing agility, reducing response times, and minimizing potential downtime through proactive interventions.

Chapter 14: Future Trends and Disruptive Innovations

Autonomous factories represent the next evolution in manufacturing automation, where AI agents and robotics work together to create self-operating production environments. These advanced manufacturing systems minimize human intervention while maximizing efficiency, precision, and adaptability. AI agents play crucial roles in decision-making processes, workflow optimization, and system coordination. However, implementing autonomous factories presents significant challenges in terms of system complexity, legacy equipment integration, cybersecurity, and ensuring scalability while maintaining reliability and safety standards.

Autonomous factories leverage AI and robotics to automate production processes with minimal human involvement, resulting in enhanced efficiency and precision across manufacturing operations. These systems excel at resource optimization, intelligently managing material usage, energy consumption, and production schedules to reduce waste and improve overall sustainability. A key advantage is their flexibility and adaptability, enabling rapid response to market changes and customer demands through reconfigurable production capabilities. This adaptability allows manufacturers to adjust production lines quickly for different products or varying demand levels.

AI agents serve three critical functions in autonomous factory operations. First, they analyze complex datasets to extract actionable insights that guide operational decisions effectively, processing information from multiple sources simultaneously. Second, these AI systems make real-time operational decisions that streamline processes and enhance manufacturing efficiency, reducing bottlenecks and

149

optimizing throughput. Third, AI agents coordinate workflows across different production stages, reducing errors and accelerating manufacturing cycles. This coordination ensures smooth transitions between processes and maintains optimal production flow throughout the entire manufacturing system.

Several challenges must be addressed for successful autonomous factory implementation. System complexity presents the primary obstacle, as these factories involve multiple interconnected components and software layers that must work seamlessly together. Legacy equipment integration requires careful planning and adaptation strategies to incorporate existing machinery with new autonomous technologies. Cybersecurity risks become critical concerns, as factories must protect sensitive data and maintain operational safety during automation deployment. Finally, ensuring scalability while maintaining reliability and safety standards is essential for successful autonomous factory deployment across different manufacturing environments.

Quantum computing represents a revolutionary technology that could dramatically enhance manufacturing AI capabilities. While still in early development stages, quantum computing offers unprecedented computational power for solving complex manufacturing optimization problems. The technology's potential applications include supply chain optimization, advanced material design, and efficient production scheduling. However, significant barriers to adoption exist, including hardware limitations, high error rates, and a shortage of quantum computing expertise in the manufacturing sector.

Chapter 14: Future Trends and Disruptive Innovations

Understanding quantum computing's manufacturing relevance requires grasping three key concepts. Quantum bits, or qubits, represent information differently than classical binary bits, enabling states that provide enhanced computational power for complex calculations. The superposition principle allows qubits to exist in multiple states simultaneously, exponentially increasing computational possibilities and enabling parallel processing of vast solution spaces. These quantum properties enable specialized algorithms that can optimize manufacturing processes more effectively than classical computers, particularly for complex problems involving multiple variables, constraints, and optimization targets.

Quantum computing promises significant breakthroughs in three key manufacturing areas. Quantum algorithms can revolutionize supply chain management by optimizing complex logistics networks and significantly reducing delays through superior route planning and inventory management. In material design optimization, quantum computing can accelerate the development of advanced materials with optimal properties by simulating molecular interactions more accurately than classical computers. Additionally, quantum algorithms can streamline scheduling processes across manufacturing operations, improving production speed while reducing operational costs through more efficient resource allocation and timeline optimization.

Several barriers currently limit quantum computing adoption in manufacturing. Hardware limitations present the most significant challenge, as current quantum systems face stability and scalability issues that affect practical applications. High error rates in quantum computations hinder reliable processing and require sophisticated error correction methods that are still under development. The shortage of quantum computing specialists limits widespread adoption and development within the manufacturing industry. However, the future outlook remains promising, with advancing hybrid quantum-classical research indicating potential integration into AI-driven manufacturing systems within the coming decade.

AI-driven circular manufacturing systems represent a sustainable approach to industrial production that aligns with environmental and economic goals. These systems focus on creating closed-loop processes where materials are continuously reused, waste is minimized, and product lifecycles are extended. AI plays a crucial role in optimizing resource flows, predicting material needs, and enhancing recycling processes. This approach not only reduces environmental impact but also creates economic value through improved resource efficiency and waste reduction strategies.

Circular manufacturing operates on three fundamental principles that transform traditional linear production models. Closing the loop involves creating systems where materials are continuously reused and waste is minimized, moving away from the traditional take-make-dispose approach. Extending product lifecycles requires redesigning processes to maximize product longevity while encouraging repair, refurbishment, and reuse rather than replacement. Reducing environmental impact through sustainable methods involves minimizing resource consumption and pollution while supporting environmentally friendly industrial growth that balances economic and ecological objectives.

AI enables circular manufacturing through three key optimization strategies. Resource flow analysis uses AI to examine complex material flows, identifying inefficiencies and opportunities to optimize material use throughout manufacturing processes. Material needs prediction leverages AI algorithms to forecast future material requirements accurately, streamlining supply chains and minimizing excess inventory that could become waste. Recycling process optimization employs AI to enhance material recovery rates and reduce production waste by improving sorting accuracy, processing efficiency, and identifying opportunities for material reuse within production cycles.

Real-world applications demonstrate AI's transformative impact on circular manufacturing. Material recovery innovation shows how AI enhances recovery processes, reducing waste and improving recycling efficiency across various manufacturing industries through better sorting and processing techniques. Energy efficiency improvements result from AI-driven solutions that optimize energy consumption patterns, lowering both operational costs and carbon footprints in industrial operations. Supply chain redesign benefits from AI-enabled optimization that enhances resource utilization and accelerates circular economy adoption by identifying opportunities for material reuse and waste reduction throughout the supply network.

This chapter has covered four transformative AI innovations reshaping manufacturing. Foundation models accelerate innovation through versatile AI frameworks that enhance manufacturing processes and decision-making capabilities. Autonomous systems enable self-operating machinery that increases efficiency while reducing human error and operational costs. Quantum computing promises to solve complex manufacturing optimization problems faster than classical computers, potentially revolutionizing how we approach industrial challenges. Finally, circular manufacturing focuses on sustainability by minimizing waste and promoting resource reuse, creating both environmental and economic benefits for forward-thinking manufacturers.

www.ingramcontent.com/pod-product-compliance
Lightning Source LLC
Chambersburg PA
CBHW060034210326
41520CB00009B/1129